大学英语特色课程系列教材

College English for Textile and Fashion Majors (Book One)

纺织服装专业
大学英语综合教程（第1册）

总主编　胡伟华
主　编　梁满玲　副主编　邹　妮　杜国香　赵月阳　麻　瑞

国家一级出版社
全国百佳图书出版单位

图书在版编目（CIP）数据

纺织服装专业大学英语综合教程 : College English for Textile and Fashion Majors (Book One). 第1册 / 梁满玲主编 ; 胡伟华总主编. -- 西安 : 西安交通大学出版社, 2021.9

ISBN 978-7-5693-2270-5

Ⅰ. ①纺… Ⅱ. ①梁… ②胡… Ⅲ. ①纺织工业－英语－高等学校－教材②服装工业－英语－高等学校－教材 Ⅳ. ①TS1②TS941

中国版本图书馆CIP数据核字(2021)第185764号

纺织服装专业大学英语综合教程（第1册）
College English for Textile and Fashion Majors (Book One)

总 主 编　胡伟华
主　　编　梁满玲
策划编辑　蔡乐芊
责任编辑　蔡乐芊　庞钧颖
责任校对　张　娟
封面设计　任加盟

出版发行　西安交通大学出版社
（西安市兴庆南路1号　邮政编码710048）
网　　址　http://www.xjtupress.com
电　　话　（029）82668357　82667874（市场营销中心)
（029）82668315（总编办）
传　　真　（029）82668280
印　　刷　陕西思维印务有限公司

开　　本　850mm×1168mm　1/16　印张　20.5　字数　485千字
版次印次　2021年9月第1版　2021年9月第1次印刷
书　　号　ISBN 978-7-5693-2270-5
定　　价　59.90元

如发现印装质量问题，请与本社市场营销中心联系调换。
订购热线：（029）82665248　（029）82665249
投稿热线：（029）82665371
读者信箱：xjtu_rw@163.com

前　言
Preface

一、编写背景

本教材以习近平新时代中国特色社会主义思想为指导，是深入贯彻落实全国教育大会和新时代全国高等学校本科教育工作会议精神的成果。本书编写团队根据《教育部关于加快建设高水平本科教育全面提高人才培养能力的意见》，以提升高校专业人才培养质量，践行大学英语教学改革为目标，在“外语+”模式背景下，开发了这套大学英语特色课程系列教材，旨在通过深化英语课程基础性和应用性的结合，服务国家纺织服装专业的人才培养工作，增强国家纺织服装产业在国际社会的话语表达能力。

我们需要培养大学生在各自专业及行业领域里汲取和交流专业信息的能力，以及用英语直接开展工作的能力，因此，亟需能够满足此类需求的相关课程和教材。掌握专业英语对于学生了解快速变化的纺织服装行业尤为重要。

基于此，本书编写团队本着严谨、务实的态度，深入调研相关用人单位和学生的需求，精心策划、编写了这套适合高校纺织服装专业学生使用的大学英语特色教材。

二、编写思路

本教材以《大学英语教学指南（2020版）》为指导，力求准确把握大学英语教学的性质、目标及要求，以及新时代国家和社会对大学生英语水平及专业能力的需求，在设计和编写中体现人文性和工具性，力求全面提升学生的语言水平和综合素养。同时，本教材借鉴近年来大学英语教学改革的成功经验与教学实践的成果，希望通过新的教学思路及设计进一步推动大学英语教学迈向新台阶。

总体来说，本教材在设计和编写中遵循以下原则：

1. 教学目标体现个性化教学的需求。

2. 教学理念体现“教师主导，学生主体”。

3. 教学内容反映时代特色，聚焦热点话题。

4. 教学手段体现数字化和立体化。

三、教材特色

本教材参考借鉴国内外优秀教材，基于混合式教学的实践和学生学情分析编写，具有以下特色。

1. 选材新颖，体现时代和专业特色

本教材主要面向纺织服装相关专业的本科学生，帮助学生在语言习得的同时获取本专业最前沿的知识和信息。鉴于此，本教材所选文章均具有一定的专业特色，同时紧跟时代潮流，力求最大限度地拓展学生的国际视野。这是本教材编写的一个立足点和出发点，也是区别于其他一般教材的特色所在。

2. 思政统领，落实立德树人的根本任务

本教材内容体现思政主题，力求培养学生的家国情怀。具体来讲，本书所设 8 个单元中，每个单元均有反映"中华文化走出去"的内容，例如中国的传统丝绸、中国的品牌、中国的人工智能等，引导学生树立文化自信。

3. 紧密结合大学生学习与生活，趣味性强

美国语言学家克拉申教授在 20 世纪 80 年代初期提出"语言输入说"(Input Hypothesis)，其核心内容之一就是学生二语习得的材料要遵循"既有趣又相关"的原则，即输入的语言材料越有趣越关联，语言习得的效果越好。因此，本书编写团队在编写前对相关专业学生进行了深入的调研，最终结合教材的内在逻辑和完整性拟定了每个单元的主题。教材所选内容非常贴合大学生的兴趣点，对学生自主学习具有一定的促进意义。

4. 训练学生思辨能力

"新文科"要求打破学科专业壁垒，运用多样化的教育组织方式，抓好教学"新基建"。为更好地培养学生的思辨能力，本教材结合教学实践，在每单元都设置了相应的思辨类话题，题目类型包括思维导图、问卷调查、课堂演讲，课堂辩论等。这些话题紧密结合本单元主题，以多样化的形式丰富了本单元内容，帮助学生拓宽思路，增强思辨能力。

5. 混合式教学模式

混合式教学模式是在多种学习理论的指导下，根据教学内容、学生和教师自身条件，混合传统的面授和网络传授两种课堂形式，以达到预期的学习目标的一种教学模式。本教材提供了丰富的线上线下资料，为混合式教学提供了充分条件，具有很强的实操性。

四、使用建议

本教材旨在有效巩固和扎实培养学生的语言知识技能和专业素养，通过视角多元、内涵丰富、与时俱进的选材以及形式多样的练习，帮助夯实学生的专业词汇基础，提升学生的专业英语阅读能力，培养学生的国际交流能力。根据专门用途英语阶段的基本要求本教材共设计 2 册，每册有 8 个单元，分别供两个学年使用，各部分具体内容如下：

1. Pre-Reading Activity

此部分由视听材料和听说练习引出主题，激发学生对单元主题的兴趣及思考，有效促进学生对单元主题内容的理解。

2. In-Depth Reading

此部分为单元主体内容，由 Text A 和 Text B 两个部分组成。两篇文章围绕单元主题，各包含一篇课文和相关练习。单元主题以纺织服装行业的发展历程为主线，涵盖历史、材料工艺、文化、杰出人物、工艺发展、时尚潮流、品牌故事及智能服饰。文章选材既扎根本土文化又放眼世界，符合纺织服装专业学生的专业和兴趣，从价值引领、技能培养，以及开阔视野等多个维度，加强学生跨文化思辨意识，满足学生思想、情感与学习的需求。

3. Exercises

练习部分是针对 Text A 和 Text B 两篇文章设计的语言技能练习，由以下内容构成：

（1）Reading Comprehension：此题在 Text A 和 Text B 两篇课文后均有设置，包括课文理解和思辨问题，考查学生对课文主旨、重要细节、文章内涵的理解，同时也通过开放式问题及训练，培养学生思辨与创新能力。

（2）Language Enhancement：此题在 Text A 和 Text B 两篇课文后均有设置，旨在通过词汇练习加强学生对相关专业词汇的理解和运用，帮助学生学以致用，将语言学习和纺织服装专业知识有效结合起来。

（3）Sentence Structure：此题只在 Text B 课文后设置，主要针对文中出现的重点句型，通过补全翻译和句子改写的练习，帮助学生掌握语法使用规范，进一步夯实语言基础。

（4）Translation：此题只在 Text A 课文后设置，包含英译汉与汉译英两篇段落翻译，内容围绕单元主题，将专业内容进行拓展，在训练学生翻译能力的同时，引导学生理解和表达中西方在纺织服装产业方面存在的差异，提高学生的专业跨文化交际能力。

（5）Writing：该训练只在 Text A 课文后设置，内容从段落写作逐级进阶到篇章写作，从结构、方法和润色等方面培养学生的写作能力。

4. Extensive Reading

该部分文章内容同样与单元主题相关，学生通过自主阅读，训练英语阅读能力，进一步拓宽专业视野。

本系列教材总主编是西安工程大学胡伟华教授。第一册主编为梁满玲，参加编写的主要人员有梁满玲、邹妮、杜国香、赵月阳、麻瑞。本书的编写也得到西安工程大学人文社会科学学院的大力支持，在此表示由衷感谢！同时对本书参考文献的著作者、出版社编辑和各位工作人员一并表示诚挚谢意！

由于时间仓促，加之编者水平有限，不足之处敬请读者批评指正。

编者

2021 年 4 月

目　录
Contents

Unit 1　History of Textile and Clothing 1

Unit 2　Textile Fibers 39

Unit 3　Textile Craft and Technology 77

Unit 4　Prominent Figures 115

Unit 5　Western Costume and Culture 157

Unit 6　Fashion and Trend 195

Unit 7　Brand Stories 233

Unit 8　Artificial Intelligence and Fashion 275

Glossary 308

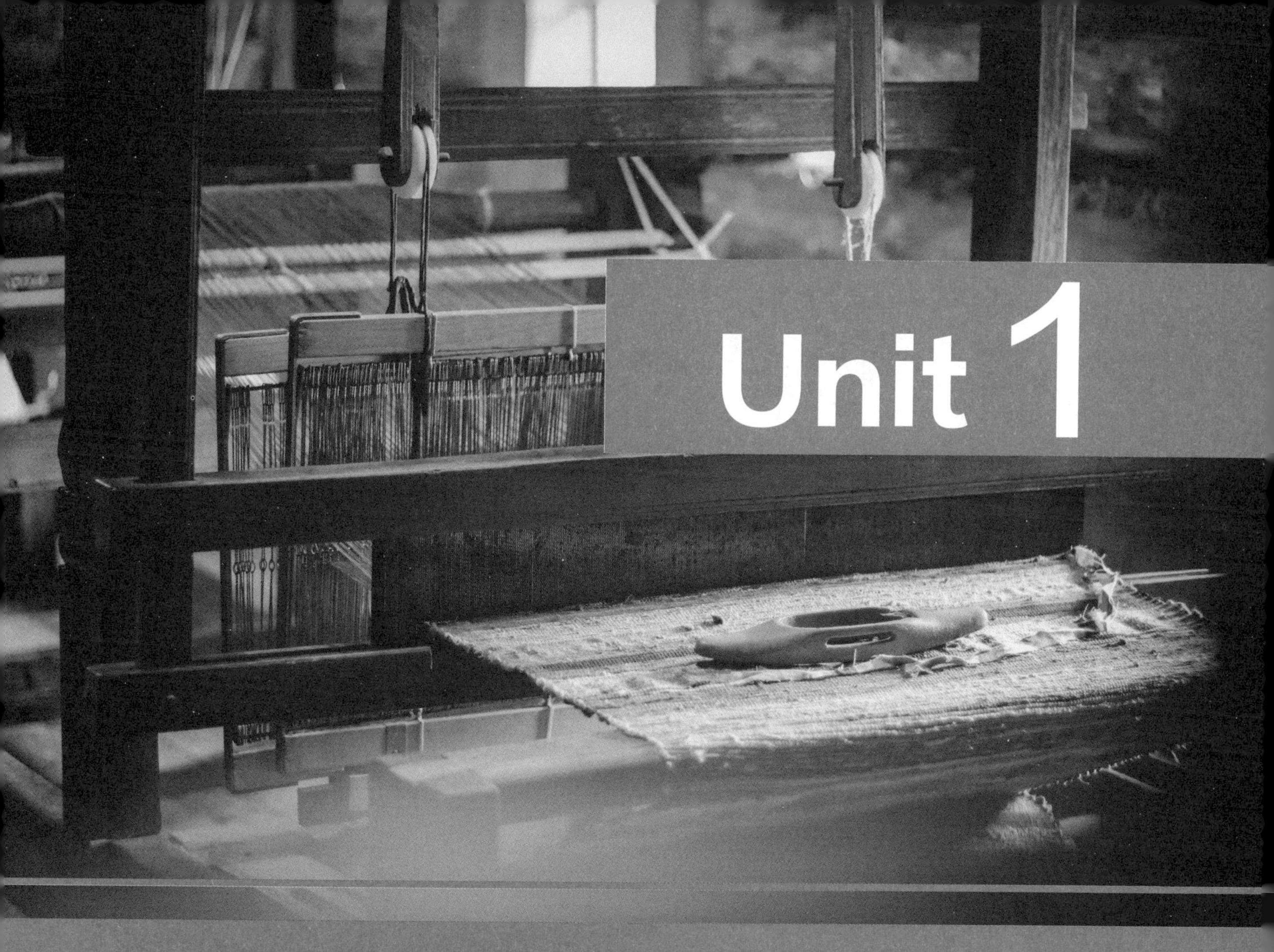

Unit 1

History of Textile and Clothing

The story of humanity is the story of textiles—as old as civilization itself.

—*Virginia Postrel*

Vain trifles as they seem, clothes have, they say, more important offices than to merely keep us warm. They change our view of the world and the world's view of us.

—*Virginia Woolf*

Pre-Reading Activities

1. Watch the video and choose the best answer to each of the following questions.

(1) Why did the speaker find the same intimacy and warmth from the show?

A. Because she had a very close relation with her grandmother.

B. Because her grandmother taught her how to stitch and knit when she was a child.

C. Because she loved sewing.

(2) Why did the show touch the speaker so much?

A. Because she missed her grandmother.

B. Because the weaver was very wise.

C. Because it reminded her that weaving as an "early technology" was enduring.

2. Listen to the talk again and fill in the blanks based on what you hear.

We opened up the back of the (1) ________, so you can have that cool blue light, coming in behind the (2) ________, and then you have the warm theatre (3) ________and they meet in the middle. The world of (4)________ and the world of the theatre also meet in the world of the (5) ________. It's the (6) ________ of the things, so it's unmaking and it's making and it's the sound of the bell and it's the silence of the theatre, the darkness, the (7) ________. It's the way thread right becomes smoke. It's how all of those make a (8)________.

3. Discuss the following questions with your partner.

(1) Do you have any experience of weaving, sewing or knitting? If so, how do you feel when you do it?

(2) Do you think it is necessary for schools to teach students such traditional skills? And why?

Text A

The History of Weaving and the Textile Industry

1 Have you ever stopped to wonder how the clothes you're wearing right now, that much-loved T-shirt or new pair of jeans, were made? Most of us don't consider the **intricacies** of the textile industry, but the history of clothing and cloth production **spins** a rich and colourful tale that should be in everyone's **repertoire**. With that in mind, we're taking it back to basics by shedding some light on the history of weaving and textiles—just to give you a little something to mull over the next time you load up your wash bag with your favourite **outfits**.

Ancient Weaving

2 To understand the practice of weaving and its role in the **thriving** textile industry, we need to follow the threads of this ancient art all the way back to **prehistory** ...

3 To give an **incredibly** brief and slightly dry summary that in no way does the revolutionary process justice, the art of weaving involves **entwining** a set of **vertical** threads, the "**wrap**", with a set of **horizontal** threads, the "**weft**". The practice itself seems to be almost **ingrained** in human nature, because even before the process of weaving was implemented, its **underlying** principles were applied in the creation of everyday necessities like shelters and baskets.

4 These crafts relied upon the **interlacing** of small materials, such as **twigs** and leaves, to form stable objects. Once ancient humans discovered how to entwine plant **fibres** to create thread some 20 or 30 thousand years ago, these basic weaving principles were put to extensive use and **elaborate**

and highly practical items were **manufactured** through the art of finger weaving, a skill still widely practised today.

5 Weaving itself is one of the oldest surviving practices in the world, with a history rooted in the Neolithic Period (c. 9000–4000 BCE). It was at this time that the creation of woven fabrics exploded, with every household producing cloth for personal use. Weaving became an **indispensable** skill for Neolithic people and was consequently closely connected to the family unit, a tradition that would **endure** for **millennia**.

Spinning and Weaving in the Middle Ages

6 The art of weaving was slowly perfected and **refined** over thousands of years, eventually leading to highly specialised cloth produced by skilled practitioners. It is no surprise that the production of this cloth, demanding higher levels of skill, coincided with the gradual movement of weaving away from the household and into the workplace. By the Middle Ages, a well-developed supply chain consisting of **dyers**, spinners, weavers, **fullers**, **drapers**, and tailors had been implemented to support the **booming** textile and weaving industry that was fast becoming one of the most **lucrative** trades across Europe. The city of Coventry was made particularly wealthy through the explosive weaving trade. Such was the city's fame that the saying, "true blue", is alleged to have descended from the longer phrase, "as true as Coventry blue", in reference to the city's **knack** for producing blue dyes that didn't run and thus remained "true".

7 At this time, weaving in Europe continued to take place at the loom that had **dominated** the weaving process for millennia, although a number of improvements, imported from China and other global empires, were gradually introduced to **expedite** the process. For instance, in the 11th-century the introduction of horizontal, foot-operated looms enabled an easier, much more efficient weaving process. Furthermore, the spinning wheel, likely originating in India sometime between 500 CE and 1000 CE and eventually imported to Europe from the Middle East, replaced the

earlier method of hand spinning. Far more than a mere **staple** of the fairy tale tradition, the spinning wheel continued to be improved until it was able to greatly expedite the process of turning fibres into **yarn** in preparation for weaving. The resulting yarn shortage **underscored** the necessity of mechanising the process, paving the way for the explosive advancements that were to occur throughout the Industrial Revolution.

Weaving in the Industrial Revolution

8 In 1774, a heavy tax on cotton thread and cloth made in Britain was **repealed**, likely sparked by a number of revolutionary developments within the trade. The inventions that sparked these developments included the Flying Shuttle (1733), which allowed wider cloth to be woven at a faster speed than previously possible, the Spinning Jenny (1765), which increased the number of threads a single machine could spin from six to eighty, and the Water Frame (1769), which used water as a source of power while producing a better thread than the Spinning Jenny. Samuel Crompton's Spinning Mule, developed in 1779, built upon these ideas by combining the most positive aspects of the Spinning Jenny and the Water Frame to produce the best spinning results of the age. By the 1790s, steam engines were being widely utilised in cotton factories to further improve textile production by reducing **dependency** on water, largely **negating** previous issues of water scarcity as a result.

9 These advances coincided with the spread of chemical **bleaches** and dyes, enabling bleaching, dyeing and printing to take place in the same location. Finally, with the invention of Robert's Power Loom in 1812, all stages of cotton making were **consolidated** and able to occur in the one factory.

10 The advances were such that the wealth of the textile industry rose rapidly throughout the mid-1700s to the mid-1800s. As a result, it quickly became the main industry of the Industrial Revolution regarding employment and invested **capital**, and was even the first to use modern production methods.

Weaving and the Textiles Industry Today

11 Today, weaving has been almost exclusively **commercialised**, although many communities and individuals around the world continue to weave by hand, either for fun, for cultural **identification**, or out of necessity. Automatic power operated looms now dominate the trade, greatly improving and **streamlining** this important aspect of the textile industry.

12 Although the practice of weaving has moved almost entirely out of the public eye, it remains a **crucial** step in the long supply chain **embedded** within the global fashion industry. With a history that dates back some 30,000 years, weaving is truly one of the oldest **extant** skills practised by humans on a global scale, and it is this impressive **credential** that **renders** it so deserving of a little **acknowledgement** the next time you reach for your favourite outfit!

Notes

Neolithic Period It refers to the last stage of the Stone Age; it is a term coined in the late 19th century by scholars which covers three different periods: Palaeolithic, Mesolithic, and Neolithic. The Neolithic Period is significant for its megalithic architecture, the spread of agricultural practices, and the use of polished stone tools.

Coventry 考文垂，英国城市

Flying Shuttle 飞梭

Spinning Jenny 珍妮纺纱机

Water Frame 水力纺纱机

Samuel Crompton 塞缪尔·克伦普顿

Spinning Mule 走锭细纱机，又名骡机

Power Loom 动力织布机

Industrial Revolution 工业革命，产业革命（指18世纪至19世纪机器取代人力、工业迅速发展的阶段）

New words and phrases

intricacy /'ɪntrɪkəsi/

n. ① [C] (intricacies) (pl.) **the ~ of sth** the complicated parts or details of sth 错综复杂的事物
② [U] the fact of having complicated parts, details or patterns 纷乱；复杂

spin /spɪn/

v. ~ **sth (round)** to make sth turn round and round rapidly 使某物快速旋转

repertoire /'repətwɑː(r)/

n. [C] all the plays, songs, pieces of music, etc. which a company, actor, musician, etc. knows and is prepared to perform 某一艺术团体、演员、音乐家等可演出的节目

outfit /'autfɪt/

n. [C] ① all the equipment or articles needed for a particular purpose; kit 全套装备；全套工具；全部用品 ② a set of clothes worn together (esp. for a particular occasion or purpose) 一套衣服（尤指用于某场合的衣服）

thrive /θraɪv/

v. grow or develop well and vigorously; prosper 茁壮成长，蓬勃发展；繁荣

prehistory /ˌpriː'histəri/

n. [U] the period of time in history before information was written down 史前时期

incredibly /ɪn'kredɪbli/

ad. ① extremely 极端地；极其
incredibly lucky/stupid/difficult/beautiful 极其幸运／愚蠢／困难／美丽
② in a way that is very difficult to believe 难以置信地

entwine /ɪn'twaɪn/

v. ① make sth by twisting one thing around another 编制（将一物缠绕在另一物上制成某物）
② (usually passive) **be entwined (with sth)** to be very closely involved or connected 纠缠

vertical /'vəːtɪkəl/	*a.* pointing up in a line that forms and angle of 90 degree with a flat surface 垂直的；直立的
wrap /ræp/	*v.* ~ **sth (up) (in sth)** to cover sth completely in paper or other material 包裹
horizontal /ˌhɔrɪ'zɔntl/	*a.* parallel to the horizon; flat; level 与地平线平行的；平的；水平的
weft /weft/	*n.* [sing.] (in weaving) threads taken crosswise over and under the lengthwise threads of the warp（纺织）纬纱，纬线
ingrain /in'grein/	*v.* to fix deeply or indelibly, as in the mind 牢牢记住；在心中根深蒂固地记住 *a.* made of predyed fibres; thoroughly dyed 染色的；由预先染色的纤维制成的；完全染过色的
underlie /ˌʌndə'laɪ/	*v.* ① lie or exist beneath (sth) 位于或存在于（某物）之下：the underlying clay, rock, etc. 处于下层的黏土、岩石等． ② [no passive] (formal) to form the basis of (sb's actions, a theory, etc.); account for 构成（某人行动及理论等）的基础；作（某事物）的说明或解释
interlace /ˌɪntə'leɪs/	*v.* ~ **(sth) (with sth) (cause things to)** be joined by weaving or lacing together; to cross (one thing with another) as if woven（使东西）编结，交织；使（物与另一物）交错
twig /twɪg/	*n.* [C] a small thin branch that grows out of a larger branch on a shrub or tree 细枝；嫩枝
fibre /'faɪbə/	*n.* [C] any of the slender threads of which many animal and plant tissues are formed（动植物的）纤维

elaborate /ɪ'læbərɪt/ — *a.* very detailed and complicated; carefully prepared and finished 详尽而复杂的；精心制作的
v. to describe or explain sth in detail 详尽解释或说明某事；阐述

manufacture /ˌmænjʊ'fæktʃə/ — *v.* to make (goods) on a large scale, using machinery 用机器大量制造（货物）
n. (pl.) manufactured goods or articles 制造品；产品

indispensable /ˌɪndɪ'spensəbəl/ — *a.* essential; too important to be without 必需的，必不可少的

endure /ɪn'djuə/ — *v.* ① to suffer or undergo (sth painful or uncomfortable) patiently 忍受；忍耐 ② to continue in existence 持续；持久

millennium /mi'leniəm/ — *n.* (*pl.* millennia) a span of one thousand years 一千年

refine /rɪ'faɪn/ — *v.* ① to remove impurities from (sth); purify 从（某物）中除去杂质；精制；精炼；提纯 ② to improve (sth) by removing defects and attending to detail（去粗取精、一丝不苟）改良（某事物） ③ make (sb/sth) more cultured or elegant; remove what is coarse or vulgar from... 使（某人 / 某事物）更有教养或更文雅；使改掉粗俗言行

dye /'daiˌə/ — *n.* someone whose job is dying cloth or other material 染色工

fuller /'fulər/ — *n.* [C] one that fulls cloth 漂洗工

draper /'dreɪpə/ — *n.* [C] a shopkeeper who sells cloth and clothing 布商；服装商

boom /buːm/ — *v.* to have a period of rapid economic growth 经济迅速发展
n. a. sudden increase (in population, trade, etc); a period of prosperity（人口、贸易等的）突然增加；繁荣昌盛时期

lucrative /'luːkrətɪv/ — *a.* producing much money; profitable 赚钱的；可获利的

knack /næk/ — *n.* [C] a skill at performing some special task; ability 技巧；诀窍

dominate /'dɔmɪneɪt/ — *v.* ① have control of or a very strong influence on (people, events, etc.) 支配，统治，控制，影响（人、事等）② be the most obvious or important person or thing in (sth) 在（某事物）中处于优势或占上风；占最重要地位

expedite /'ekspɪdaɪt/ — *v.* (*formal*) to accelerate the progress of (work, business, etc.); hasten or speed up 有助于（工作、业务等的）进展；加快；加速

staple /'steɪpl/ — *a.* main or principal; standard 主要的；基本的；标准的；
n. [C] sth that is produced by a country and is important for its eeonomy（其国的）主要产品，支柱产品

yarn /jɑːn/ — *n.* [U] fibres (esp. of wool) that have been spun for knitting, weaving, etc. 纱；纱线

underscore /ˌʌndə'skɔː/ — *v.* ① to draw a line under a word, sentence, etc. 在单词或句子下划线 ② to emphasise or show that sth is important or true 强调或指出重要性、真实性

repeal /rɪ'piːl/ — *v.* withdraw (a law, etc.) officially; revoke 废止（法规等）；撤销

dependency /dɪ'pendənsɪ/ *n.* [C] a country governed or controlled by another 附属国；附属地

negate /nɪ'geɪt/ *v.* ① (*formal*) to deny or disprove the existence of (sb/sth) 否定或否认（某人/某事物）的存在 ② cancel the effect of (sth); nullify 消除（某事物）的作用；使无效

bleach /bliːtʃ/ *v.* (cause sth to) become white or pale (by chemical action or sunlight) （使某物通过化学作用或日照）变白；漂白

n. [U, C] substance or process that bleaches or sterilises 漂白或消毒；漂白剂

consolidate /kən'sɔlɪdeɪt/ *v.* ① (cause sth to) become more solid, secure, or strong 使某事物巩固，加固，加强 ② (cause things to) unite or combine (into one)（使事物）联合或合并

capital /'kæpɪtl/ *n.* ① [U] wealth or property that may be used to produce more wealth 资本；资金 ② [sing.] sum of money used to start a business 本钱

commercialise /kə'məːʃəlaɪz/ (AmE also -ize) *v.* ① [usually passive] to be more concerned with making money from something than about its quality, especially in a way that other people do not approve of（尤指不择手段地）牟利

identification /aɪˌdentɪfɪ'keɪʃən/ *n.* [U, C] (*abbr.* ID) the process of showing, proving or recognizing who or what sb/sth is 鉴定；认出；身份证明

streamline /'striːmlaɪn/ *v.* ① to give a streamlined form to (sth) 使（某物）成流线型

② to make (sth) more efficient and effective, eg. by improving or simplifying working methods 使（某事物）效率更高或作用更大

crucial /'kruːʃəl/	*a.* ~ **(to/for sth)** very important; decisive 至关重要的；决定性的
embed /im'bed/	*v.* ① [usually passive] ~ **sth (in sth)** to fix sth firmly into a substance or solid object 镶嵌 ② if ideas, attitudes, or feelings, etc. are embedded, you believe or feel them very strongly 思想、态度、感觉根深蒂固
extant /ɪk'stænt/	*a.* (esp. of documents) still in existence （尤指文件）仍然存在的，现存的
credential /kri'denʃəl/	*n.* [C] something that gives a title to credit or confidence 凭据；资质
render /'rendə(r)/	*v.* ① (*formal*) ~ **sth (to sb/sth); ~ (sb) sth** to give sth in return or exchange, or as sth which is due 给予某物作为报偿或用以交换；回报；归还 ② to cause (sb/sth) to be in a certain condition 使（某人或某事物）处于某种状况
acknowledgement /ək'nɑlɪdʒmənt/	*n.* ① [sing., U] an act of accepting that sth exists or is true, or that sth is there 承认，确认② [C, U] an act or a statement expressing thanks to sb; sth that is given to sb as thanks 感谢；谢礼
shed light on	to clarify or supply additional information on 阐明，使…… 清楚地显示出
mull over	think about sth for a long time before deciding what to do 仔细考虑
load up	to put a large quantity of things or peope onto or into sth 把大量 …… 装上，装入
follow the threads of	follow the way a story develops 跟着故事发展的思路
coincide with	two events happen at the same time 同时发生

be alleged to	stated or described to be such 被指控做过某事
descend from	to have developed from sth that existed in the past 由 …… 传下来
pave the way for	to make it possible for sb to do sth or for sth to happen 为 …… 铺平道路，做好准备

Reading Comprehension

Understanding the text

Answer the following questions.

1. Why does the author believe the practice of weaving "seems to be almost ingrained in human nature"?
2. What kind of material did the ancient humans use to form stable objects?
3. When did every household start to produce cloth for personal use?
4. How did the textile and weaving industry become one of the most lucrative trades across Europe?
5. What does "as true as Coventry blue" mean?
6. What has promoted the mechanisation of weaving industry?
7. How were the steam engines widely utilised in cotton factory to improve textile production?
8. What kind of looms now dominate the weaving and textile industry?

Critical thinking

Work in pairs and discuss the following questions.

1. Since weaving has been commercialised, hand sewing is replaced by highly automated assembly lines. What's your opinion toward this change?
2. What will weaving and the textile industry be like in the future?

Research project

Marc Andreessen, co-founder of Netscape, once said "The story of technology is a story of human ingenuity, and nowhere is this clearer than the story of textiles." How do you understand this statement? Try to find related stories to demonstrate the influence of textile on human civilization.

Language Enhancement

Words in use

Fill in the blanks with the words given below. Change the form when necessary. Each word can be used only once.

fibre	boom	refine	repeal	ingrain
render	intricacy	expedite	indispensable	outfit

1. Rose explained the ________ of the job.
2. She bought a new ________ for her daughter's wedding.
3. The blow to his head was strong enough to ________ him unconscious.
4. A system that had been ________ for generations could not be easily undone by change from the top.
5. Eating cereals and fruit will give you plenty of ________ in your diet.
6. This book is ________ to anyone interested in space exploration.
7. Reading good books helps to ________ one's speech.
8. The insurance business suffered from a vicious cycle of ________ and bust.
9. The committee does not have the power to ________ the ban.
10. Aid workers are trying to ________ the process of returning refugees to their homes.

Banked cloze

Fill in the blanks by selecting suitable words from the word bank. You may not use any of the words more than once.

A. luxury	F. millennium	K. revolution
B. spin	G. monetary	L. cotton
C. imperial	H. compensation	M. cultivation
D. commercial	I. valuable	N. manufacturing
E. protein	J. convenient	O. basic

Silk is a natural 1. ________ produced by mulberry silkworm which is used for textile 2. ________. Silk fiber has a triangular prism-like structure which allows silk cloth to refract incoming light at different angles and with that to produce different colours.

History of silk began in the 27th century BCE in China where it remained in sole use until the 3. ________ ways appeared from China to the Mediterranean Sea. There is also evidence of silk dating between 4000 BCE and 3000 BCE. During the latter half of the first 4. ________ BCE, Silk Road opens and silk starts to spread the world. Cultivation of silk spread to Japan somewhere around 300 CE while by 522 CE the Byzantines managed to obtain silkworm eggs and were able to begin silkworm 5. ________ of their own.

In China, only women farmed silk worms. Many women were employed on the farms of silkworms. Silk was considered a 6. ________ item and silk became very popular among high society. Popularity was such that laws were made to regulate and limit use of silk to the members of the 7. ________ family. That rule stayed in power for over millennia.

Silk was not used just for clothing. Paper was also made out of silk and it was the first type of luxury paper. Again its worth became more 8. ________ and it was used as pay for government officials and 9. ________ to citizens who were particularly worthy. The length of the silk cloth became a 10. ________ standard in China.

Expressions in use

Fill in the blanks with the expressions given below. Change the form when necessary. Each expression can be used only once.

load up	be alleged to	descend from	coincide with
shed light on	mull over	pave the way for	follow the threads of

1. She is developing new theories that might ________ these unusual phenomena.
2. It's a fine offer, but we need time to ________ it ________.
3. Customers ________ their suitcases as usual, with no special packaging needed.
4. When you ________ these proposals, you will find that the theme is all about equal opportunity.
5. The heroic age of bridge construction ________ the expansion of the railroads.
6. She ________ have stolen more than $50,000 over the course of several years.
7. Recent evidence supports the theory that birds ________ dinosaurs.
8. The discovery ________ for the development of effective new treatments.

Translation

I. Translate the following paragraph into Chinese.

Knitting is a technique of making fabric with yarn on two or more needles. The word "knitting" comes from the word "knot". Knitting is made of wool, silk, and other fibres. The first pieces of clothing made in the technique similar to knitting were socks. These socks were done with Nålebinding which is a technique that uses single needle and thread. There are ideas that knitting began in Middle East, from there came to Europe through Mediterranean trade routes and then to the Americas from Europe. The oldest knitted items have been found in Egypt and are dated between the 11th and 14th centuries.

II. Translate the following paragraph into English.

时装设计的起源可以追溯到1826年。查尔斯·弗雷德里克·沃思(Charles Frederick Worth)被认为是世界上第一位时装设计师。他从前是一名布商，在巴黎开了一家时装店，开创了时装店的传统，并告诉顾客什么样的衣服适合他们。在他的影响下，许多设计公司开始聘请艺术家来设计服装图案，他们将衣服纸样发给客户，客户根据自己的喜好下单。

Paragraph Writing

How to Plan a Paragraph

A paragraph is composed of multiple sentences focused on a single, clearly-defined *topic*. The sentences of the paragraph explain the writer's main idea (the most important idea) about the topic. There should be exactly one main idea per paragraph, so whenever you move on to a new idea, you should start a new paragraph. In academic writing, a paragraph is often between five and ten sentences long, but it can be longer or shorter, depending on the topic.

Paragraphs are the building blocks of a larger body of essay. The way you write a paragraph will determine the quality of essay. So, before you start your essay writing, you need to learn how to write a good paragraph. Basically, a paragraph is a collection of sentences that all relate to one central topic. Before you begin writing a paragraph, you must plan what you are going to write about. The following steps will help you make a good paragraph plan.

Step 1: Choosing and narrowing a topic

Each paragraph has only one topic. The topic can not be too narrow (limited, brief) or too broad (general). A narrow topic will not have enough ideas to write about. For example, "the ages of my pets" is too narrow.You can't write very much about it. A broad topic will have too many ideas for just one paragraph. On the other hand, "pets" is too general. There are thousands of things you could say about it.

You can narrow this topic by choosing one aspect of pets to discuss. For example, "keeping pets is beneficial to the patients who have mental health problems".

Step 2: Gathering ideas

After you choose a topic, you need to brainstorm some ideas to write about in your paragraph. This can be achieved by asking questions, discussing with others, making a list and mapping the ideas. Don't worry now about whether the ideas are good or bad, useful or not. You can decide that later. Right now, you are gathering as many ideas as you can.

Step 3: Editing the ideas

After you have gathered plenty of ideas, you need to go back and edit them. This is the time to choose which ideas are the most interesting, and which are the most relevant to (important or necessary for) your topic. Of course, you can still add new ideas if you think of something else while you are re-reading your list.

The following sample will show you how a paragraph is planned.

Choosing and narrowing the topic

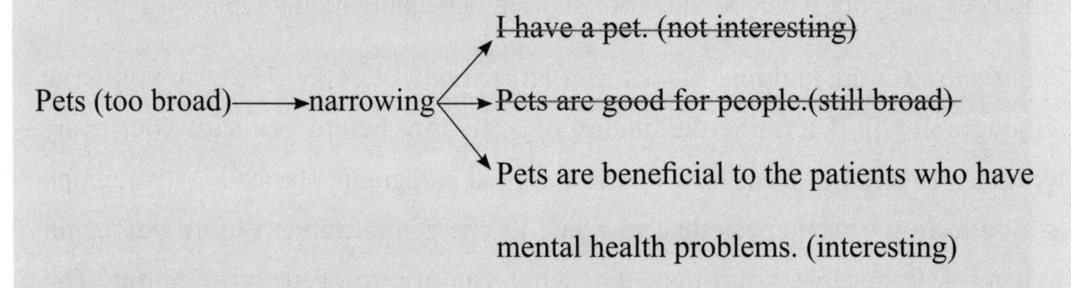

Gathering ideas

1. Why are pets beneficial?
2. What do pets do?
3. Pets can make the patients exercise through petting or playing with them.
4. Pets can encourage love.
5. Pets can accompany the patients.
6. Keeping pets needs the sense of responsibility.
7. Keeping pets is not expensive.

8. Keeping pets protects patients from having side effects of pills.
9. Pets can keep the patients safe.

Editing the ideas

Firstly, pets have many physical advantages for those with mental problems. Pets can make the patients exercise through petting or playing with them.

Furthermore, pets provide emotional healing for mental strugglers. Pets can encourage the patients to show their love and sense of responsibility.

Finally, they enable people who suffer from mental issues to be treated without any negative aspects. It won't cost you too much money to keep pets and it doesn't have any side effects like pills do.

In short, it is advantageous to use animals to treat mental health problems.

Work with a partner or in a small group. Choose and narrow one of the following topics. Then gather as many ideas as you can in five minutes.

1. Teenager fashion
2. The Industrial Revolution
3. Silk

Edit the ideas alone you have gathered.

When did Humans Start Wearing Clothes?

1 **Determining** exactly when humans began wearing clothes is a challenge, largely because early clothes would have been things like animal **hides**, which don't **fossilise** and degrade rapidly. So we cannot get **archaeological**

evidence that can be used to determine the time when our early human—"hominin"—**ancestors** stopped wandering about naked, and started **draping** their bodies with animal furs and skins.

2 Instead, **anthropologists** largely rely on indirect methods to date the origin of clothes. There have been several different theories based on what archaeologists have been able to find. For instance, based on **genetic** skin-**colouration** research, humans lost body hair around one million years ago—an ideal time to start wearing clothes for warmth. The first tools used to **scrape** hides date back to 780,000 years ago, but animal hides served other uses, such as providing **shelter**, and it's thought that those tools were used to prepare hides for that, rather than clothing. Eyed needles started appearing around 40,000 years ago, and those tools point to more complex clothing, meaning clothes had probably already been around for a while.

3 All that being said, scientists have started gathering **alternative** data that might help solve the mystery of when we humans started covering our body.

4 A recent University of Florida study concluded that humans started wearing clothes some 170,000 years ago, lining up with the end of the second-to-last Ice Age. How did they figure that date out? By studying the **evolution** of **lice**.

5 Researchers found that head lice and lice that live in our clothes separated at around this time. They observed that clothing lice are, well, extremely well-adapted to clothing. So they **hypothesised** that body lice must have **evolved** to live in clothing, which meant that they weren't around before humans started wearing clothes. The study used DNA **sequencing** of lice to calculate when clothing lice started to genetically split from head lice.

6 The findings of the study are significant because they show that clothes appeared some 70,000 years before humans started to **migrate** north from Africa into cooler climates. The invention of clothing was probably one factor that made **migration** possible.

7 This timing also makes sense due to known climate factors in that era. As Ian Gilligan, a lecturer at the Australian National University, said that the study gave "an unexpectedly early date for clothing, much earlier than the earliest solid archaeological evidence, but it makes sense. It means modern humans probably started wearing clothes on a regular basis to keep warm when they were first **exposed** to Ice Age conditions."

8 As to when humans moved on from animal hides and into **textiles**, the first **fabric** is thought to have been an early ancestor of felt. From there, early humans took up **weaving** some 27,000 years ago, based on impressions of baskets and textiles on clay. Around 25,000 years ago, the first Venus **figurines**, little **statues** of women, appeared wearing a **variety** of different clothes that pointed to weaving technology being in place by this time.

9 From there, more recent ancient **civilisations** discovered many materials they could **fashion** into clothing. For instance, Ancient Egyptians produced **linen** around 5500 BCE, while the Chinese likely started producing silk around 4000 BCE.

10 As for clothing for fashion, instead of just keeping warm, it is thought that this **occurred** relatively early on. The first example of dyed **flax** fibres was found in a cave in the Republic of Georgia and dates back to 36,000 years ago. That being said, while they may have added colour, early clothes seem to have been much simpler than the clothing we wear today—mostly cloth draped over the shoulder and pinned at the waist.

11 Around the mid-1300s in certain regions of the world, with some **technological** advances in **previous** century, clothing fashion began to change **drastically** from what it was before. For instance, clothing started to be made to form fit the human body, with **curved seams**, **laces**, and buttons. Contrasting colours and fabrics also became popular in England. From this time, fashion in the West began to change at an alarming rate, largely based on **aesthetics**, whereas in other cultures fashion typically changed only with great political **upheaval**, meaning changes came more

slowly in most other cultures.

12 The Industrial Revolution, of course, had a huge impact on the clothing industry. Before the revolution, textiles were made by hand in the "cottage industry," where materials would be brought to homes and picked up when the products were finished. New technological machinery such as Hargreaves's "spinning jenny," Richard Arkwright's water frame, and the Boulton and Watt steam engine **emerged** and improved the quality of thread and the speed it took to produce. Meanwhile, technology **contributed** to an extremely rapid fall in the price of clothing and an **enormous** increase in the **scale** of clothing **manufacturing**. The mass production of clothing began roughly in the mid-nineteenth century, when some **manufacturers** began to produce clothes that did not require fitting **session** with the **sewers** or tailors.

13 Clothes could now be made in mass in factories rather than just in the home and could be transported from factory to market in record time. As a result, clothes became drastically cheaper, leading to people having significantly larger **wardrobes** and contributing to the constant change in fashion that we still see today.

Notes

The University of Florida The University of Florida, referred to as UF, also known as UFL, is a well-known top public research university located in Gainesville, Florida, USA. It is a member of the Association of American Universities. The establishment of the university can be traced back to 1853. The University of Florida is known as the public Ivy League and ranks among the top public universities in the United States.

The study of lice Principal investigator David Reed, associate curator of mammals at the Florida Museum of Natural History on the UF campus, studied lice in modern humans to better understand human evolution and migration patterns. His study used DNA sequencing to calculate when clothing lice first began to diverge genetically from human head lice. The study was funded by the

National Science Foundation and its findings appeared in the Oct. 5, 2004, online issue of the *Public Library of Science Journal, PLOS Biology.*

Ice Age 冰期；冰川期；冰河时代

Ian Gilligan 伊恩·吉利根

Venus 金星；太白星；维纳斯（爱与美的女神）

The Republic of Georgia 格鲁吉亚共和国

New words and phrases

determine /dɪˈtɜːmɪn/ — *v.* ① to discover the facts about sth; to calculate sth exactly 查明；测定；准确算出 ② cause (sth) to occur in a particular way; be the decisive factor in 造成；在……中起决定因素 ③ firmly decide 下定决心；决意

hide /haɪd/ — *n.* an animal's skin, especially when it is bought or sold or used for leather（尤指买卖或用作皮革的）皮；毛皮

v. to go somewhere where you hope you will not be seen or found 躲避；隐匿

fossilise /ˈfɒsəlaɪz/ — *v.* ① [usually passive] to become or make sth become a fossil（使）变成化石，石化② to become, or make sb/sth become, fixed and unable to change or develop（使人或物）僵化

archaeological /ˌɑːkiəˈlɒdʒɪkl/ — *a.* related to archaeology, or carried out for the purposes of archaeology 考古学(上)的

ancestor /ˈænsestə(r)/ — *n.* ① a person in your family who lived a long time ago 祖宗；祖先 ② an early type of animal or plant which others have evolved（动植物的）原始种型

drape /dreɪp/

v. ① **~ sth around/over/across, etc. sth** to hang clothes, materials, etc. loosely on sb/sth 将（衣服、织物等）悬挂，披；② **~ sb/sth in/with sth** to cover or decorate sb/sth with material 遮盖；盖住；装饰

anthropologist /ˌænθrəˈpɒlədʒɪst/

n. a person who studies anthropology 人类学家

genetic /dʒəˈnetɪk/

a. connected with genes or genetics 基因的；遗传学的

colouration /ˌkʌləˈreɪʃn/

n. the natural colours and patterns on a plant or an animal（植物或动物的）自然色彩，自然花纹

scrape /skreɪp/

v. ① to remove sth from a surface by moving sth sharp and hard like a knife across it 除去；刮掉；削去 ② to rub sth by accident so that it gets damaged or hurt 损坏；擦坏；擦伤；刮坏；蹭破

shelter /ˈʃeltə(r)/

n. [U] ① the fact of having a place to live or stay, considered as a basic human need 居所；住处

② **~ (from sth)** protection from rain, danger or attack 遮蔽，庇护，避难（避雨、躲避危险或攻击）

alternative /ɔːlˈtɜːnətɪv/

n. a thing that you can choose to do or have out of two or more possibilities 可供选择的事物

a. ① (of one or more things) available as another possibility 可供替代的；另外的 ② different from the usual or traditional way in which sth is done 非传统的；另类的

evolution /ˌiːvəˈluːʃn/

n. ① the gradual development of plants, animals, etc. over many years as they adapt to changes in their environment 进 化 ② the gradual development of sth in a particular situation or over a period of time 演变；发展；渐进

lice /laɪs/

n. (plural of louse) small insects that live on people's skin and in their hair 虱子

hypothesise /haɪˈpɒθəsaɪz/

v. to suggest a way of explaining sth when you do not definitely know about it; to form a hypothesis 假设；假定

sequencing /ˈsiːkwənsɪŋ/

n. [U] gene sequencing or DNA sequencing involves identifying the order in which the elements making up a particular gene are combined (对整套基因、分子程序的)排序

split /splɪt/

v. ① to divide, or to make sth divide, into two or more parts 分开，使分开(成为几个部分) ② to tear, or to make sth tear, along a straight line (使)撕裂

migrate /maɪˈgreɪt/

v. ① to move from one town, country, etc. to go and live and/or work in another 移居；迁移 ② to move from one part of the world to another according to the season(随季节变化)迁徙

migration /maɪˈgreɪʃn/

n. the movement of large numbers of people, birds or animals from one place to another 迁移；移居；迁徙

expose /ɪkˈspəʊz/

v. ① make something visible, typically by uncovering it 使暴露；使显露；使露出② to tell the true facts about a person or a situation, and show them/it to be immoral, illegal, etc. 揭露；揭穿

textile /ˈtekstaɪl/	*n.* any type of cloth made by weaving or knitting 纺织品
fabric /ˈfæbrɪk/	*n.* ① material made by weaving wool, cotton, silk, etc., used for making clothes, curtains, etc. and for covering furniture 织物；布料 ② **the ~ (of sth)** the basic structure of a building, such as the walls, floor and roof（建筑物的）结构（如墙、地面、屋顶）
felt /felt/	*n.* a type of soft thick cloth made from wool or hair that has been pressed tightly together 毛毡
weave /wiːv/	*v.* ① to make cloth, a carpet, a basket, etc. by crossing threads or strips across, over and under each other by hand or on a machine called a loom（用手或机器）编，织 ② **~ sth (out of/from ...) | ~ sth (into...)** to make sth by twisting flowers, pieces of wood, etc. together（用……）编成
figurine /ˌfɪɡəˈriːn/	*n.* a small statue of a person or an animal used as a decorative object（人、动物的）小雕像，小塑像
statue /ˈstætʃuː/	*n.* a figure of a person or an animal in stone, metal, etc., usually the same size as in real life or larger 雕塑，雕像，塑像（大小通常等于或大于真人或实物）
variety /vəˈraɪəti/	*n.* ~ **(of sth)** the quality or state of being different or diverse; the absence of uniformity, sameness, or monotony 多样性；变化，多变性
civilization /ˌsɪvəlaɪˈzeɪʃn/	*n.* [U] ① a state of human society that is very developed and organized 文明 ② all the people in the world and the societies they live in, considered as a whole 文明世界；文明社会

fashion /ˈfæʃn/	*n.* ① a popular style of clothes, hair, etc. at a particular time or place; the state of being popular（衣服、发式等的）流行款式，时兴式样 ② [U] the business of making or selling clothes in new and different styles 时装业 *v.* to make or shape sth, especially with your hands（尤指用手工）制作；使成形；塑造
linen /ˈlɪnɪn/	*n.* a type of cloth made from flax, used to make high-quality clothes, sheets, etc. 亚麻布
occur /əˈkɜː(r)/	*v.* ① to happen; take place 发生 ② to exist or be found in a place 存在；出现
flax /flæks/	*n.* a plant with blue flowers, grown for its stem that is used to make thread and its seeds that are used to make linseed oil 亚麻
technological /ˌteknəˈlɒdʒɪkəl/	*a.* relating to or using technology 与技术有关的，技术的；工艺的
previous /ˈpriːviəs/	*a.* happening or existing before the event or object that you are talking about 先前的；以往的
drastically /ˈdræstɪkli/	*ad.* in a way that is likely to have a strong or far-reaching effect 激烈地；彻底地
curve /kəːv/	*n.* a line or surface that bends gradually; a smooth bend 曲线；弧线；曲面；弯曲
seam /siːm/	*n.* a line of stitches which joins two pieces of cloth together 衣物等的缝合线；接缝
lace /leɪs/	*n.* a fine open fabric, typically one of cotton or silk made by looping, twisting or knitting thread in patterns and used especially for trimming garments 花边

aesthetic /iːsˈθetɪk/	*a.* ① concerned with beauty and art and the understanding of beautiful things 审美的；有审美观点的；美学的 ② made in an artistic way and beautiful to look at 美的；艺术的
upheaval /ʌpˈhiːvl/	*n.* a violent or sudden change or disruption to something 剧变；激变；动乱；动荡
emerge /ɪˈməːdʒ/	*v.* move out of or away from something and come into view 浮现；出现；出来
contribute /kənˈtrɪbjuːt/	*v.* ① ~ **(to sth)** help to cause or bring about; to be one of the causes of sth 促成；促使；有助于 ② to give sth, especially money or goods, to help sb/sth 捐助；捐款
enormous /ɪˈnɔːməs/	*a.* extremely large in size, quantity, or extent 巨大的；庞大的；极大的
scale /skeɪl/	*n.* ① the size or extent of something 范围 ② an indication of the relationship between the distances on a map and the corresponding actual distances 比例尺
manufacturer /ˌmænjuˈfæktʃərə(r)/	*n.* a person or company that produces goods in large quantities 生产者；制造者；生产商
session /ˈseʃn/	*n.* a period of time devoted to a particular activity 一段时间；阶段
sewer /ˈsjuːə(r)/	*n.* ① a person or thing that sews 缝纫工；缝纫机 ② an underground pipe that is used to carry sewage away from houses, factories, etc. 污水管；下水道；阴沟
wardrobe /ˈwɔːdrəʊb/	*n.* a large closet or freestanding cupboard for hanging clothes 衣橱

date back to	to have been made in or to have come into being in (a certain time in the past) 追溯到；从…… 开始有
point to	① to direct attention to (sb or sth) by moving one's finger or an object held in one's hand in a particular direction 指向 ② to mention or refer to (sth) as a way of supporting an argument or claim 表明
line up with	① to join someone or something in a line 与……对齐 ② to be in agreement or accordance with something 一致
figure out	discover a way or solve a problem 解决；算出；想出；理解；断定
make sense	to have a clear meaning and be easy to understand; be reasonable 有意义；讲得通；言之有理
due to	as a result of 由于；应归于
on a regular basis	quite often and/or in a consistent, regular manner 经常地；定期地
as to	as for; about; according to 至于，关于；就……而论
take up	to become interested in a particular activity or subject and spend time doing it 拿起；开始从事；占据（时间或地方）
a variety of	a lot of particular types of things that are different from each other 种种；各种各样的
be in place	① in the correct position 在适当的位置 ② existing and ready to be used 就位；准备好

as for	on the topic of; regarding; as to; with regard to 关于，至于
at a/an...rate	in the speed of 以……的速度
have a/an...impact on	have a/an... influence on 对……有……影响
contribute to	to help to cause sth happen 有助于；带来，促成

Reading Comprehension

Understanding the text

Choose the best answer to each of the following questions.

1. Why is it challenging to determine when humans began wearing clothes?
   A. Because we can't find archaeological evidence.
   B. Because people don't care about it.
   C. Because opinions of scientists vary from one to another.
   D. Because it is mysterious.
2. According to the archaeological research, what did the appearance of eyed needles indicate?
   A. It indicated the origin of clothes.
   B. It indicated that clothes had probably already been around for a while.
   C. It indicated that humans got smarter.
   D. It meant that humans started using tools.
3. How did scientists figure out when clothes originated?
   A. By analysing textile fibres.
   B. By researching genetic colouration of human being.
   C. By studying the evolution of lice.
   D. By analysing archaeological evidence.
4. According to the text, when did humans move on from animal hides and into textiles?
   A. 70,000 years ago.

B. 27,000 years ago.

C. 36,000 years ago.

D. 25,000 years ago.

5. What impact did the industrial revolution have on the clothing industry?

A. It made fashion come into being.

B. It made clothes more sophisticated.

C. It took the place of traditional hand-made techniques.

D. It made mass production of clothing possible.

Mind mapping

The following mind-map presents you an overview of the process determining the origin of the first clothes. Find out the key information and then finish the map to complete the outline.

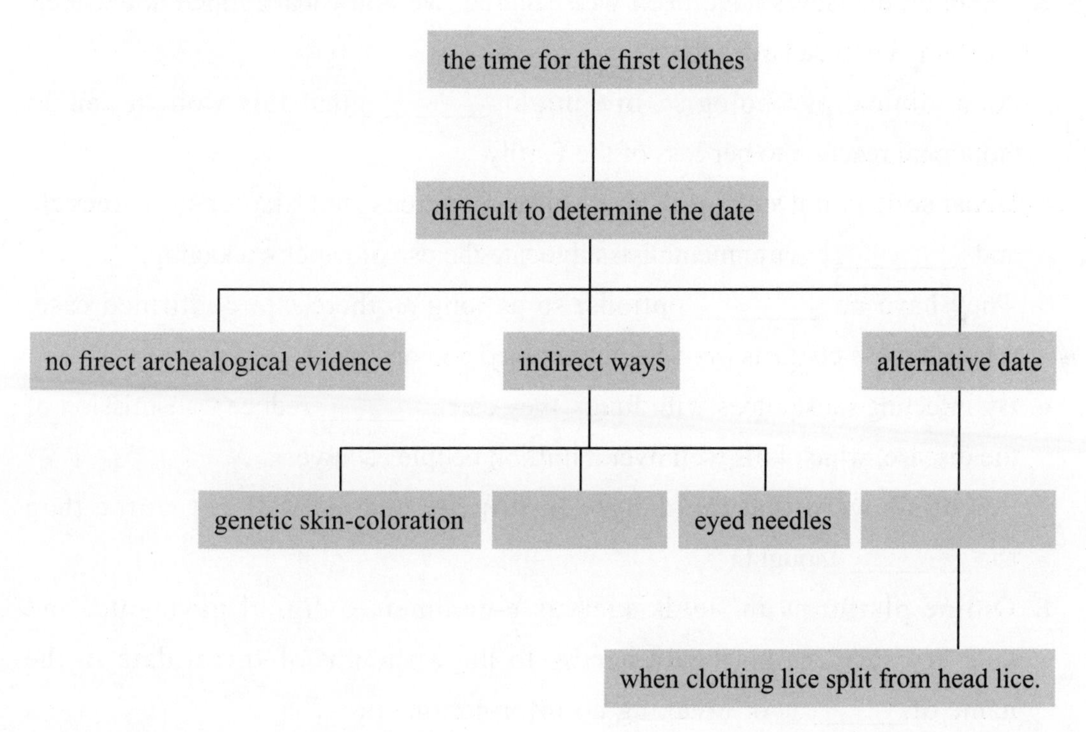

Language Enhancement

Words in use

Fill in the blanks with the words given below. Change the form when necessary. Each word can be used only once.

determine	fossilise	degrade	previous	alternative
aesthetic	drastically	migrate	innovation	hypothesise

1. As farmland shrinks, many rural people are forced to ________ to urban areas in search of work.
2. Although dinosaurs have been well studied, we don't learn much about their breathing because lungs don't ________ .
3. As a clinical psychologist, one might ________ that this woman had an emotional reaction to her loss of the family.
4. Because it is not only rich in natural resources, but also easy to recycle and ________, environmentalists advocate the use of paper packaging.
5. They have no ________ options, so as long as there is a confirmed case, relevant close contacts must be quarantined and observed.
6. By infecting mosquitoes with fungi, they can ________ reduce transmission of the disease, which kills well over a million people each year.
7. Scientists warn that the danger of climate change will get worse than the ________ thought.
8. Online platforms in fields such as e-commerce, digital payments, and delivery services may gain access to huge amount of social data in the name of ________ or breaking up information silo.
9. Acquired cultivation is one of the ________ factors in the development of a person's character.
10. Clothing culture, which has always been a symbol of human evolution and civilisation, has unique ________ value.

Expressions in use

Fill in the blanks with the expressions given below. Change the form when necessary. Each expression can be used only once.

point to	line up with	figure out	on a regular basis
due to	a variety of	be in place	contribute to

1. Unfortunately, ____________ unforeseen circumstances, this year's show has been canceled.
2. Your performance against your job description and progress made towards achieving your objective will be reviewed ____________.
3. Researchers released new data recently which ____________ an increase in technology use among children is changing the very nature of children.
4. In order to ensure a safe and smooth running route for all runners, traffic control measures will ____________ throughout the race.
5. Then write a desire and intention list. Cross-reference these lists and see if your to-do actions ____________ your desires and intention.
6. Obstacles don't have to stop you. If you run into a wall, don't turn around and give up. ____________ how to climb it, go through it, or work around it.
7. She didn't show up in class frequently, therefore, her poor class attendance helped ____________ to her poor grades.
8. All that added weight leaves them at risk for ____________ health problems, including diabetes, high blood pressure and heart disease.

Sentence structure

I. Complete the following sentences by translating the Chinese into English, using "that being said, ..." structure.

Model: We cannot get archaeological evidence that can be used to determine the time when our ancestors started wearing clothes. ____________________. (话虽如此，科学家们已经开始寻找其他数据来解开这一奥秘。) →We cannot get archaeological evidence that can be used to determine the time. All that being said, scientists have started gathering alternative data to solve this mystery.

1. All the statistics in this table are collected from the Medieval clothing manufacturers.

____.

（尽管如此，这些数据在某种程度上仍然很有价值。）

2. Quarrels between the couple are difficult to avoid. ____________________

____________________.（话虽如此，但谁输谁赢并不是婚姻生活的重点）.

3. Many people complain it is hard to lose weight successfully.______________

______________________________.（话虽如此，有些人通过简单的健身以及饮食计划，轻松实现了这个目标）.

II. Rewrite the following sentences by using "it makes/doesn't make sense to do.../that..."

Model: To Ian Gilligan, it is meaningful that the study of lice gave an unexpectedly early date for clothing.

→To Ian Gilligan, it makes sense that the study of lice gave an unexpectedly early date for clothing.

1. It is unreasonable for the husband that his wife always complains about having no clothes to wear when her wardrobe has been full.

____

____.

2. Even though it is his fault, it is of little meaning to blame him right now.

____

____.

3. This stupid robber actually robbed a plainclothes man in his last robbery! So it was reasonable that he was arrested on the spot.

____

____.

How is Silk Produced from Silkworms?

1 The importance of silk in the cloth-making industry cannot be **exaggerated**. Textiles made of silk fiber are common throughout the world, especially in China where silk production was started. Clothes made from silk are particularly popular because of their **texture**. Such material has a **shimmering** appearance and is able to **refract** the light from different angles, thus producing different colors. Silk fiber can be obtained from several insects but the most popular one is **silkmoth**. Important to note is that the fiber is only produced by the **larvae** of the insect. Silk can also be obtained from other insects such as **wasps**, bees, ants, beetle, flies, and **fleas**.

Brief History of Silk Production

2 According to historians, silk production originated in China during the **Neolithic** period of the Stone Age. Production of silk was **confined** to the region of China until the opening of Silk Road in the late half of the first **millennium** BC. However, even with the opening of the trade route, China still maintained a **monopoly** over the production of silk for thousands of years. Silk was not only used for making clothes but also used for several other **applications** such as writing. From around 300 AD, the **cultivation** of silk reached Japan and by 552 AD, the **Byzantines** had managed to **cultivate** silkworm eggs and were able to begin its **cultivation**. Around the same time, the **Arabs** began manufacturing silk. The spread of **sericulturerendered** Chinese silk export less important although they still **dominated** the **luxury** silk market. Silk found its way into Western Europe

via the **Crusades**. From the Italian states, silk was exported to the rest of Europe. Europe's silk industry was completely transformed during the Industrial Revolution. New weaving techniques improved the **efficiency** of production. However, it was during this period that the silk industry faced **stiff** competition from the cotton industry.

About the Silkworm

3 **Domestic** silkmoth, scientifically referred to as **Bombyx mori**, is a moth belonging to the family **Bombycidae**. It is an economically important insect and the chief producer of silk. Silk is mainly produced by silkworm which is the **larva** of the silkmoth. Silkworms mainly feed on **mulberry** leaves as well as other mulberry **species**. There are four main species of silkworms, namely Mulberry silkworm, Muga silkworm, Tasar silkworm, and Eri silkworm. The mulberry silkworm is divided into three groups; **univoltine** (mainly found in greater Europe) and **bivoltine** (found in China, Korea, and Japan). The univoltine type **generates** silk only once a year while the bivoltine type generates it twice a year. The third type is **polyvoltine** which is only found in the tropics and can have up to 8 separate lifecycles in a year.

Process of Producing Silk

4 The process of silk production is known as sericulture. This process is divided into several stages but typically starts from cultivating silkworms on mulberry leaves. The female silkworm lays 300-400 eggs and **insulates** them on the leaves of the mulberry tree. In about 10-14 days each of the eggs **hatches** and produces a larva which is also known as a **caterpillar**. The larvae are covered in tiny dark hair. The larva begins **molting** when the head turns darker and may molt about four times after which the body turns slightly yellow and skin becomes tighter. As the larva prepares to

enter the **pupal** stage, it **encloses** itself in a cocoon which is made of raw silk produced by the **salivarygland**. The final molt from larva to **pupa** takes place in the cocoon.

5 Harvesting of silk begins at the stage when the worm starts **pupating** in their cocoons. If the worm is left to survive after spinning its cocoon, the **enzymes** produced may be **destructive** to the silk and may cause the fiber to break down. To ensure that quality fiber is produced, the silkworm cocoons are dissolved in boiling water. The worms are killed by the heat and the hot water enables the **extraction** of long fibers. For 1 kg of silk to be produced, 3,000 silkworms must feed on 104 kg of mulberry leaves. The cocoon is made of raw silk thread of about 1,000-3,000 feet long and 0.0004 inches in **diameter**. It requires about 3,000 cocoons to produce one pound of silk.

6 Because the process of **extracting** silk from the cocoon involves the killing of the larva, sericulture has attracted **criticism** from animal rights **activists**. It is for this reason that the majority of the activists promote cotton spinning. Mahatma Gandhi is known for among other things criticizing silk production based on Ahimsa's philosophy of "not hurting any living thing."

Silkworm Domestication and Breeding

7 Unlike the wild silkworm, the **domestic** silkworm has an increased cocoon size, higher growth rate, and an efficient **digestion** system. It has also adapted to living around humans and in crowded conditions. A **domesticated** male silkworm cannot fly and needs assistance to find a mate. Since its first **domestication** about 5,000 years ago in China, the production **capacity** has increased nearly tenfold. The principle of **genetic** breeding has been applied to silkworms for **maximum** output and is now

only second to **maize** in **exploiting** the principle of crossbreeding and **heterosis**.

8 The aim of silkworm breeding is to improve output and maximize income. The main area of focus is improving the egg-laying capacity, disease resistance, health of the larva, and quality of the cocoon. It is a common understanding that a healthy larva leads to a healthy cocoon. However, the health of larva is dependent on **pupation** rate, shorter larval duration, and few dead **larvae** during molting. The healthy larva has a high pupation rate and cocoon weight.

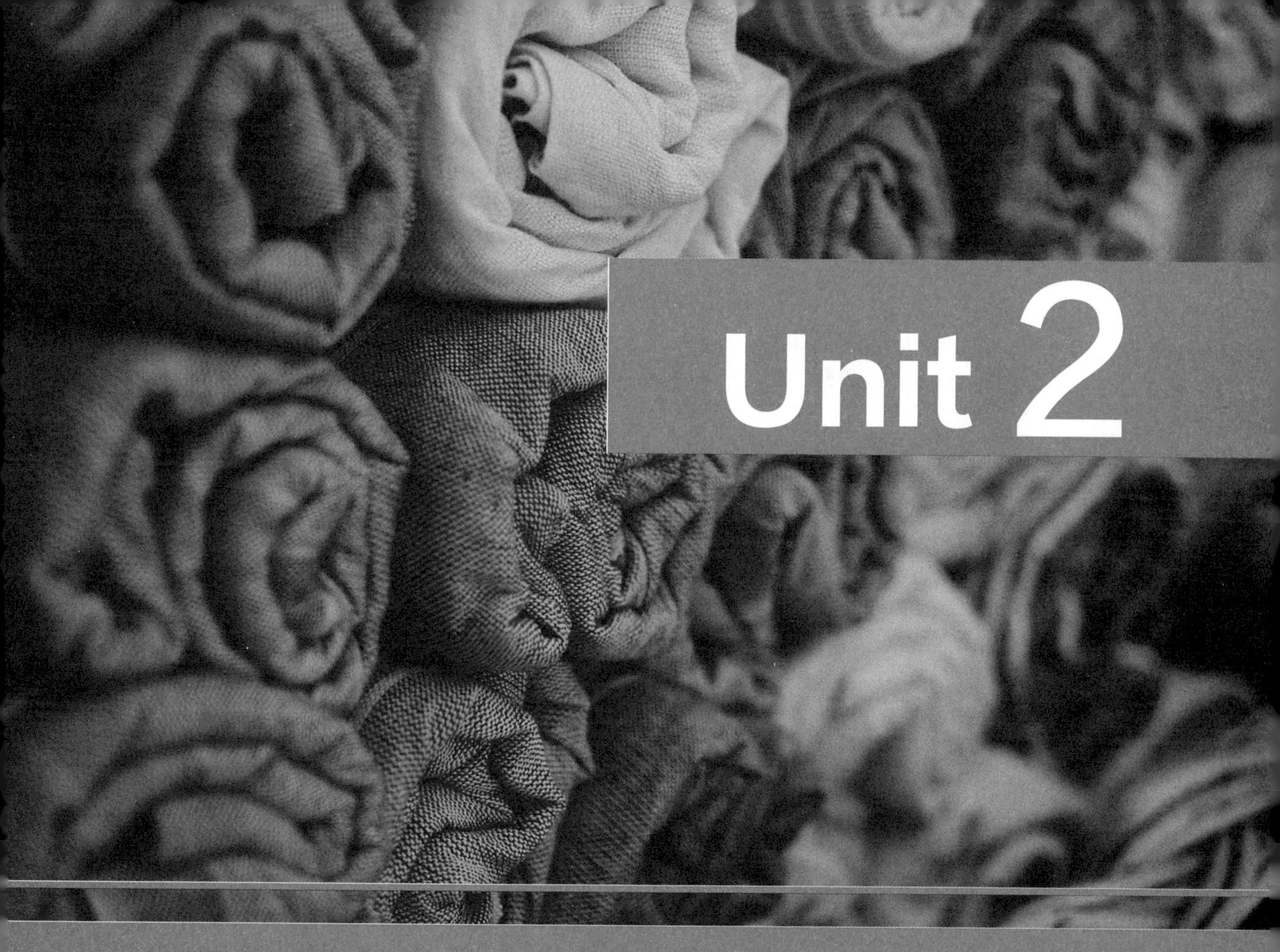

Unit 2

Textile Fibers

I grew up in the age of polyester. When I got to touch real silk, cotton and velvet, the feel of non-synthetic fabrics blew me away. I know it's important how clothing looks, but it's equally important how it feels on your skin.

— *Colleen Atwood*

Living in the midst of abundance we have the greatest difficulty in seeing that the supply of natural wealth is limited and that the constant increase of population is destined to reduce the American standard of living unless we deal more sanely with our resources.

— *Wallace Carothers*

Pre-Reading Activities

1. Listen to the book review about *Cotton, A Global History* and fill in the missing information.

(1) Mr. Beckert's answer is that for 900 years, until 1900, cotton was the world's most important ____________.

(2) In India, cotton as a cottage industry was so ____________ that it established ____________ in Britain.

(3) The success of cotton industry had two consequences. The first was ____________ in the industrial north. The second, introduced in 1774 to assist English spinners and weavers, was protectionist legislation that made it illegal to ____________.

(4) Mr. Beckert, a history professor at Harvard, calls this new economic order "____________" as it is based on ____________, expropriation of land, and ____________.

(5) Deprived of ____________ when the civil war broke out in 1861, English manufacturers rediscovered India. By 1862, ____________ of Britain's cotton originated in India.

(6) In the late 19th century, the cotton industry in England began to decline. At the height of the Great Depression in ____________ only ____________ of the world's mechanical spindles were operating in ____________, compared with ____________ in ____________. The terrible blight that has overwhelmed cotton towns such as Rochdale began then, and has grown worse since. By the late ____________ Britain accounted for only ____________ of global cotton exports.

(7) Today the main sources of raw cotton are ____________ and ____________. Mr. Beckert's story is both inspirational and utterly depressing, a ____________ of the white-knuckle ride that has been the ____________ through the centuries.

2. Watch the video about wool and answer the following questions.

(1) What is the definition of wool?

(2) What are the features of wool?

(3) How much wool does an average sheep produce every year?

(4) What is lanolin and its use?

(5) What are the procedures to make the wool into a sweater?

3. Discuss the following questions and complete the task.

(1) What kinds of textile fibers do you know? Can you give some examples about how these fibers are used in everyday life?

(2) Watch the video and get to know the basics about nylon. Then present a brief introduction of nylon in class in terms of its brief history and uses.

Text A

Textile Fibers

1 Fibers play an incredibly important part in the textile manufacturing process. Fibers used to make yarn are considered as textile fibers, which are **extracted** from different sources. But not all fibers are suitable for textiles. Textile fibers are those with **properties** that allow them to be spun into yarn or directly made into fabric, which means the fibers need to be strong enough to hold their shape, **flexible** enough to be shaped into a fabric or yarn, **elastic** enough to stretch, and **durable** enough to last.

2 Fibers are generally divided into two broad categories: natural fibers and **synthetic** or man-made fibers. Natural fibers refer to the fibers obtained from natural growing sources such as plants, animals and minerals, which have a wide range of **applications** in the manufacture of **composite** materials. Synthetic fibers are fibers in which either the basic **chemical** units have been formed by chemical **synthesis** followed by fiber

formation or the **polymers** from natural sources have been **dissolved** and **regenerated** after passage through a **spinneret** to form fibers.

NATURAL FIBERS

Animal Fibers

3 Wool, silk, and hair are typical examples of animal fibers. Wool, the fiber of the hairy cover of sheep, goats, camels, and other animals, is a **valuable** textile fiber that has high **elasticity**, **hygroscopicity**, and **thermal insulation** properties. Wool is used mainly for the production of yarn for suits, dresses, coats, **industrial** fabrics, and knit outerwear. Wool may be felted, which makes possible its use in the production of felt goods, such as felt boots, hats, and other articles.

4 The silk fiber is the only fiber obtained from an insect—**silkworm**. Silk fibers are natural **fibrous** protein-based materials, spun by **Lepidoptera larvae** such as silkworms, spiders, **scorpions**, **mites**, and flies. Raw silk consists of **filaments** obtained by unwinding **cocoons** and is used directly or after braiding, mainly in the production of fabrics for dresses and **undergarments**, as well as industrial fabrics. Recently silk is preferred for biomedical, textile and biotechnology industries due to its unique **non-toxicity**, biocompatibility, and biodegradability properties. Silk, rightly called the queen of textiles for its **lustre**, **sensuousness**, and glamour, rivals with the most advanced synthetic polymers, yet the production of silk does not require **harsh** processing conditions.

Plant/Vegetable/Cellulosic Fibers

5 Natural plant fibers are cell walls that occur in the stem, wood, and leaf parts, which are comprised of **cellulose**, hemicelluloses, **lignins**, waxes, and other **lipids**. Many useful fibers have been obtained from various parts of plants including leaves, stems (bast fibers), fruits and seeds. Plant fibers are classified into two groups: soft fibers and hard fibers.

6 The most important, widespread and inexpensive textile fiber is cotton, a

strong, thin and **hygroscopic** fiber developed from the seeds of the cotton plant. The fiber is generally transformed into yarn which is woven to manufacture fabrics. Cotton has been used for millennia in the **confection** of fabric, with the earliest known use dating from 12,000 years BCE in Egypt. Cotton yarn is used for the production of consumer fabrics for undergarments and clothing, as well as that of industrial fabrics, various knitted goods and sewing thread.

7 Bast fibers are produced from the stems, leaves and fruit of plants, usually in the form of industrial fibers. The thinnest stem fiber, **flax**, is very strong and hygroscopic, resistant to stretching. Flax yarn is used for the production of packaging, undergarments, clothing and industrial fabrics whereas hemp, a coarse stem fiber obtained from the hemp plant, is used to produce rope and coarse fabrics. The most widespread coarse stem fiber is **jute**, which is used for the preparation of bags for sugar and other materials. A new mineral-based fiber known as the basalt fiber possesses very high mechanical, chemical and thermal properties. The researchers are trying to use this fiber as an effective alternative to synthetic fibers.

8 Different factors affect the properties of natural fiber at different stages of its growth and extraction, such as plant growth, harvesting stage, fiber extraction methods, and supply chain. The favourable features of natural fibers are **sustainability**, renewability, abundance and biodegradability. Natural fibers find wide application in certain construction materials in construction industry or in the manufacture of medical biomaterials. For example, the natural fiber Chitin can be used to remove certain toxic pollutants from industrial water discharge.

SYNTHETIC FIBERS

9 Synthetic fibers are usually subdivided into two groups: regenerated and synthetic fibers. Regenerated fiber is created by dissolving the cellulose area of plant fiber in chemicals and making it into fiber again. Since it consists of cellulose like cotton and hemp, it is also called "regenerated

cellulose fiber." Truly synthetic fibers are synthesised completely from chemical substances such as petroleum-based by-products. Developed in the laboratories at the industrial scale, synthetic fibers are generally very long and possess high mechanical, thermal and physical properties, which are greatly varied according to different combinations of chemicals and fiber extraction processes. **Rayon**, nylon and **polyester** are three of the most commonly used synthetic fibers.

10 Rayon, the first man-made fiber, is an artificial textile material composed of regenerated and purified cellulose obtained from wood pulp, which was developed in the 19th century as a **substitute** for silk. Rayon is soft, **absorbent**, comfortable and easy to dye in a wide range of colours, usually mixed with cotton to make bedsheets or with wool to make carpets.

11 Sourced from coal, water and air, nylon is a **thermoplastic** silky material that can be melt-processed into fibers, films or shapes. As a tough material, nylon is difficult to tear and exhibits excellent **abrasion** resistance. Nylon fibers were the first truly synthesised man-made fibers introduced in the 1930s as a **replacement** for boar bristles in toothbrushes, which quickly found itself adapted into several forms of manufactured items. Nowadays, nylon finds its application in seat belts, sleeping bags, socks, as well as ropes for rock climbing, making parachutes and fishing nets.

12 Polyester is a synthetic, non-renewable fiber derived from coal, water, air and petroleum, which has **surpassed** cotton as the most commonly produced fiber. They are used in varying applications ranging from apparel to home textiles or can be processed for industrial applications.

13 Synthetic fibers display many advantages over natural fibers. For instance, synthetic fibers are more durable as they are mostly unaffected by living organisms and microorganisms. Moreover, many synthetic fibers offer consumer-friendly functions such as stretching, waterproofing and stain resistance. Nevertheless, the availability of the petrochemical resources is rapidly reducing and their reserves are uncertain in nature.

More importantly, using these petrochemicals yields a lot of pollution and damages the environment. Synthetic fibers that are produced from petroleum, like nylon and polyester, are not biodegradable.

14 Considering all these factors, scientists had taken up the task of replacing these synthetic fibers with the natural fibers, which has led to the research on natural fiber-based composites or bio-composites. The strength of the natural fibers has encouraged materials scientists to bring a lot of natural fibers as **reinforcements** into the composite materials.

Notes

Lepidoptera larvae 鳞翅目幼虫

New words and phrases

extract /ɪkˈstrækt/ *v.* to remove or obtain a substance from sth, for example by using an industrial or a chemical process 提取；提炼
n. [U, C] a substance that has been obtained from sth else using a particular process 提取物；浓缩物；精；汁

property /ˈprɒpəti/ *n.* ① [C, usually pl.] a quality or characteristic that sth has 性质；特性 ② [U] a thing or things that are owned by sb; a possession or possessions 所有物；财产；财物

flexible /ˈfleksəbl/ *a.* ① able to bend easily without breaking 柔韧的；可弯曲的；有弹性的 ② able to change to suit new conditions or situations 能适应新情况的；灵活的；可变动的

elastic /ɪˈlæstɪk/ *a.* ① able to stretch and return to its original size and shape 有弹性的；有弹力的 ② that can change or be changed 灵活的；可改变的；可伸缩的
n. [U] material made with rubber, that can stretch and then return to its original size 橡皮圈（或带）；松紧带

durable /ˈdjʊərəbl/

a. likely to last for a long time without breaking or getting weaker 耐用的；持久的

synthetic /sɪnˈθetɪk/

a. artificial; made by combining chemical substances rather than being produced naturally by plants or animals 人造的；（人工）合成的

n. [C] an artificial substance or material 合成物；合成纤维（织物）；合成剂

application /ˌæplɪˈkeɪʃn/

n. [U, C] ① ~ **(of sth) (to sth)** the practical use of sth, especially a theory, discovery, etc.（尤指理论、发现等的）应用，运用 ② ~ **(to sb) (for sth/to do sth)** a formal (often written) request for sth, such as a job, permission to do sth or a place at a college or university 申请；请求；申请书；申请表

composite /ˈkɒmpəzɪt/

n. [C] something made by putting together different parts or materials 合成物；混合物；复合材料

a. [only before noun] made of different parts or materials 合成的；混成的；复合的

chemical /ˈkemɪkl/

a. ① connected with chemistry 与化学有关的；化学的 ② produced by or using processes which involve changes to atoms or molecules 用化学方法制造的；化学作用

n. [C] a substance obtained by or used in a chemical process 化学制品；化学品

synthesis /ˈsɪnθəsɪs/

n. (*pl.* **syntheses**) ① [U, C] ~ **(of sth)** the act of combining separate ideas, beliefs, styles, etc.; a mixture or combination of ideas, beliefs, styles, etc. 综合；结合；综合体 ② [U] (*technical*) the natural chemical production of a substance in animals and plants（物质在动植物体内的）合成 ③ [U] (*technical*) the artificial production of a substance that is present naturally in animals and plants（人工的）合成

polymer /ˈpɒlɪmə(r)/

n. [C] (*chemistry*) a natural or artificial substance consisting of large molecules that are made from combinations of small simple molecules 聚合物；多聚体

dissolve /dɪˈzɒlv/

v. ① ~ **sth (in sth)** to make a solid become part of a liquid 使（固体）溶解 ② to officially end a marriage, business agreement or parliament 解除（婚姻关系）；终止（商业协议）；解散（议会）③ ~ **(in sth)** (of a solid) to mix with a liquid and become part of it 溶解 ④ to disappear; to make sth disappear 消除（使）消失，消散

regenerate /rɪˈdʒenəreɪt/

v. ① to grow again; to make sth grow again 再生；使再生 ② to make an area, institution, etc. develop and grow strong again 使振兴；使复兴；发展壮大

spinneret /'spɪnəret/

n. [C] a multi-pored device through which a plastic polymer melt is extruded to form fibers（纺）喷丝头

valuable /ˈvæljuəbl/

a. ① ~ **(to sb/sth)** very useful or important 很有用的；很重要的；宝贵的 ② worth a lot of money 很值钱的；贵重的

elasticity /ˌiːlæˈstɪsəti/

n. [U] the quality that sth has of being able to stretch and return to its original size and shape 弹性；弹力

hygroscopicity /haɪgrəskɒˈpɪsiti/

n. [U] the property of absorbing or discharging moisture according to circumstances 吸水性，吸湿性

thermal /ˈθəːml/

a. [only before noun] ① (*technical*) connected with heat 热的；热量的 ② (of clothing) designed to keep you warm by preventing heat from escaping from the body 保暖的；防寒的 ③ (of streams, lakes, etc.) in which the water has been naturally heated by the earth 温暖的；热的

insulation /ˌɪnsjuˈleɪʃn/ *n.* [U] the act of protecting sth with a material that prevents heat, sound, electricity, etc. from passing through; the materials used for this 隔热；隔音；绝缘；隔热（或隔音、绝缘）材料

industrial /ɪnˈdʌstriəl/ *a.* [usually before noun] ① connected with industry 工业的；产业的 ② used by industries 用于工业的

silkworm /ˈsɪlkwəːm/ *n.* [C] a caterpillar that produces silk thread 蚕

fibrous /ˈfaɪbrəs/ *a.* [usually before noun] made of many fibers; looking like fibers 纤维构成的；纤维状的

scorpion /ˈskɔːpiən/ *n.* [C] a tropical animal like an insect, with a curving tail and a poisonous sting 蝎子

mite /maɪt/ *n.* [C] a very small creature like a spider that lives on plants, animals, carpets, etc. 螨

filament /ˈfɪləmənt/ *n.* [C] a long thin piece of sth that looks like a thread 细丝；丝状物

cocoon /kəˈkuːn/ *n.* [C] a covering of silk threads that some insects make to protect themselves before they become adults 茧

undergarment /ˈʌndəˌɡɑːmənt/ *n.* [C] a piece of underwear 内衣

non-toxicity /nɔn'tɔksisəti/ *n.* [U] the property of not being poisonous or harmful to your health 无毒性

lustre /ˈlʌstər/ *n.* [U] (*NAmE* luster) ① the shining quality of a surface 光泽；光辉 ② the quality of being special in a way that is exciting 荣光；光彩；荣耀

sensuousness /ˈsenʃuəsnəs/ *n.* [U] a sensuous feeling 知觉；敏感

harsh /hɑːʃ/ | *a.* ① cruel, severe and unkind 残酷的；严酷的；严厉的 ② very difficult and unpleasant to live in 恶劣的；艰苦的

cellulose /ˈseljuləʊs/ | *n.* [U] a natural substance that forms the cell walls of all plants and trees and is used in making plastics, paper, etc. 纤维素

lignin /ˈlɪgnɪn/ | *n.* [U] a complex polymer occurring in certain plant cell walls making the plant rigid 木质素

lipid /ˈlɪpɪd/ | *n.* [C, U] (*chemistry*) any of a group of natural substances which do not dissolve in water, including plant oils and steroids 脂质；类脂

hygroscopic /haɪgrə(ʊ)ˈskɒpɪk/ | *a.* (of a substance) tending to absorb water from the air 吸湿性的

confection /kənˈfekʃn/ | *n.* [C] a thing such as a building or piece of clothing, that is made in a skillful or complicated way 精工制作的物品（如建筑物或衣物）

flax /flæks/ | *n.* [U] ① threads from the stem of the flax plant, used to make linen 亚麻纤维 ② a plant with blue flowers, grown for its stem that is used to make thread and its seeds that are used to make linseed oil 亚麻

jute /dʒuːt/ | *n.* [U] (= thin threads) from a plant, also called jute, used for making rope and rough cloth fibers 黄麻纤维

sustainability /səˌsteɪnəˈbɪləti/ | *n.* [U] the property of being sustainable 持续性；永续性；能维持性

rayon /ˈreɪɒn/ | *n.* [U] a fiber made from cellulose; a smooth material made from this, used for making clothes 人造丝；人造丝织品

polyester /ˌpɒliˈestə(r)/ *n.* [U] a strong material made of fibers which are produced by chemical processes, often mixed with other materials and used especially for making clothes 聚酯纤维；涤纶

substitute /ˈsʌbstɪtjuːt/ *n.* [C] ~ (**for sb/sth**) a person or thing that you use or have instead of the one you normally use or have 代替者；代替物；代用品

absorbent /əbˈzɔːbənt; əbˈsɔːbənt/ *a.* able to take in sth easily, especially liquid 易吸收（液体等）的

thermoplastic /ˌθɜːməʊˈplæstɪk/ *a.* (*technical*) a plastic material that can be easily shaped and bent when it is heated, and that becomes hard when it is cooled 热塑（性）塑料

abrasion /əˈbreɪʒn/ *n.* ① [U] damage to a surface caused by rubbing sth very hard against it 磨损 ② [C] a damaged area of the skin where it has been rubbed against sth hard and rough（皮肤、表皮）擦伤处；（表层）磨损处

replacement /rɪˈpleɪsmənt/ *n.* [C] a thing that replaces sth, especially because the first thing is old, broken, etc. 替代品；替换物

surpass /səˈpɑːs/ *v.* (*formal*) to do or be better than sb/sth 超过；胜过；优于

reinforcement /ˌriːɪnˈfɔːsmənt/ *n.* [U, sing.] the act of making sth stronger, especially a feeling or an idea （感情或思想等的）巩固，加强，强化

be suitable for having the right qualities for a particular person, purpose, or situation 适用于；合适的

divide into to separate into two or more parts 划分

refer to to mention or speak about sb or sth 参考；涉及；指的是

obtain from to get sth that you want, especially through your own effort, skill, or work 获得；得到

a wide/broad/whole/full range of a number of people or things that are all different, but are all of the same general type 广泛的，大范围的

consist of to be formed from two or more things or people 由……组成；由……构成；包括

rival with compete with 相竞争；相比较

be comprised of to have sb/sth as parts or members 由……组成

classify into to divide what group sth belongs to according to a system 把……分类；将……归类

transform into to undergo a change or conversion into sb or sth with a markedly or drastically different appearance, form, nature of function 转变

be resistant to ① not damaged or affected by sth 对……有抵抗力的 ② opposed to sth and want to prevent it from happening 抵制的

according to ① in a way that depends on differences in situations or amounts 依据；按照；根据 ② as stated or reported by sb/sth 据（……所说）；按（……所报道）

range from … to … ① to include both the amounts from A to B and anything in between 处于某范围内，在某范围内变化 ② to include a variety of different feelings, actions, etc. 包含不同的感情或行为等

replace A with/by B to remove someone from their job or sth from its place, and put a new person or thing there 替换；调换

Reading Comprehension

Understanding the text

Answer the following questions.

1. According to the first paragraph, what properties do textile fibers have?
2. What is the general classification of textile fibers? What is the definition for each category?
3. What are the characteristics of silk as a preferred textile fiber?
4. What is cotton yarn mainly used for?
5. What are the favorable features and the application of natural fibers?
6. When and how were nylon fibers first introduced and applied?
7. What are the advantages of synthetic fibers over natural fibers?
8. What challenges do synthetic fibers present and what may be the possible solutions in textile production?

Critical thinking

Work in pairs and complete the following tasks.

1. In your opinion, what are the differences between natural fibers and synthetic fibers?
2. In addition to the fibers mentioned in the text, there are many more kinds of textile fibers, such as bamboo fiber, carbon fiber and glass fiber. Please choose one of them and give a brief introduction to it.

Mind mapping

The following mind map will present you with an overview of the classifications of commonly used textile fibers. Please complete the outline with what you have learned in this text.

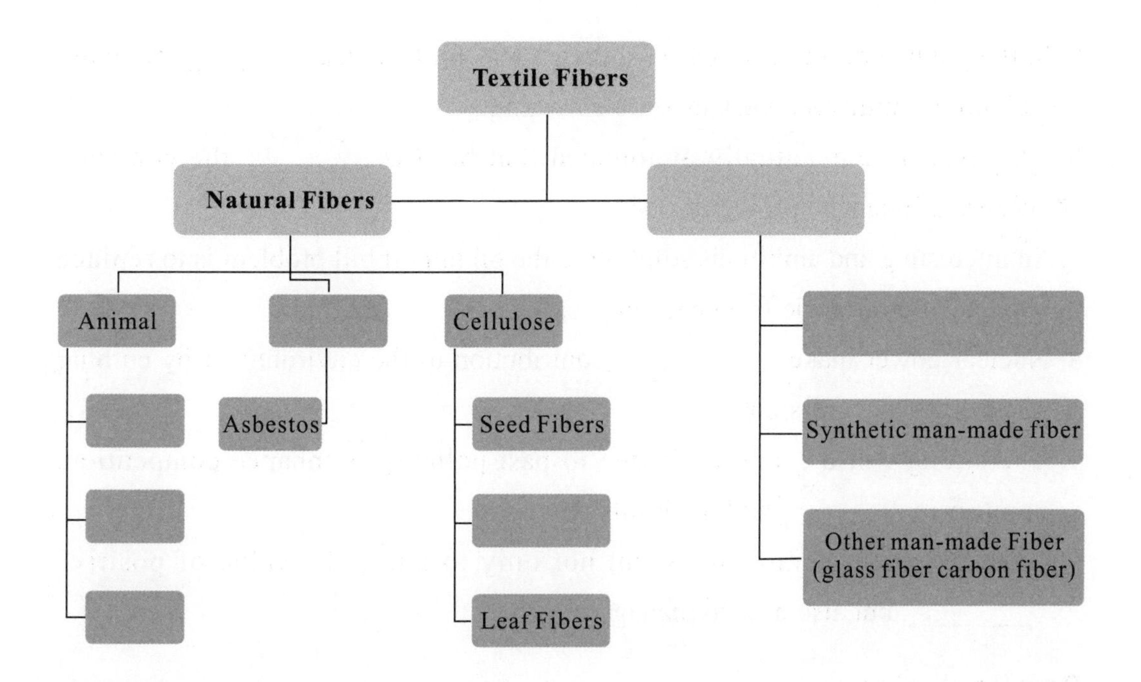

Language Enhancement

Words in use

Fill in the blanks with the words given below. Change the form when necessary. Each word can be used only once.

flexible	application	valuable	substitute	reinforcement
synthetic	property	dissolve	sustainability	surpass

1. Yet the permanent achievements of her reign were not ________ by any other ruler of the age.
2. Writing is downgraded as if it is a clumsy ________ for more efficient means of preserving data.
3. The researchers will test the chemical and biological ________ of the samples to make sure the findings of the experiment are reliable.
4. We need a ________ management system, able to meet the changing needs of our customers.

5. Initially much of the Unix-based software was for technical ________ but more and more commercial packages are emerging.

6. The executive eventually announced that he was ________ the company because of financial problems.

7. An interesting and ambitious solution to the oil import bill problem is to replace ________ petrol made from natural gas.

8. Nuclear power makes a ________ contribution to the environment by curbing carbon dioxide emissions.

9. The strong world economy is due to past policies to enhance competition, economic ________ and flexibility.

10. The excellent companies seem not only to know the value of positive ________, but also how to manage it.

Banked cloze

Fill in the blanks by selecting suitable words from the word bank. You may not use any of the words more than once.

A. natural	F. implications	K. recycled
B. considerate	G. performance	L. diversified
C. consumption	H. expected	M. applications
D. maintainable	I. considerable	N. maximize
E. minimize	J. sustainable	O. synthetic

When most people think of textiles, they think about traditional textiles meant for clothing or home furnishings. However, the use of textiles is considerably more 1. ________and hi-tech than what most assume. Non-traditional or technical 2. ________ of textiles account for nearly 27% of the global textile market. The technical textile sector is counted to be one of the fastest growing sectors.

Technical textiles are materials that focus more on 3. ________ rather than appearance and are currently in demand in multiple industries. According to a report by *Fortune Business Insights* that analysed the sector from 2019 to 2026, the market value of technical textiles was valued at $159.29 billion in 2018 and is 4. ________ to grow at a CAGR of 2.7% between 2010 and 2026. The sector consists of both natural and 5. ________ fibers but the latter holds the largest revenue share in the global technical textiles market. Some of the synthetic fibers used in this industry are nylon, polyester, acrylic, olefin, PLA and modacrylic, all of which are produced from petrochemicals and emit 6. ________ amounts of carbon dioxide. To ensure their commitment towards preservation of the planet and provide us with 7. ________ solutions for the textile industry, many textile companies have developed fibers from 8. ________ materials and natural sources for technical textiles in recent years.

The 9. ________ of man-made fibers in technical industries is very high, which results in increase of greenhouse gases. Time has come to focus more on eco-friendly fibers. The use of eco-fibers and recycled fibers are the best solutions to keep our environment clean and 10. ________ global warming.

Expressions in use

Fill in the blanks with the expressions given below. Change the form when necessary. Each expression can be used only once.

resistant to	consist of	derive from	be suitable for
according to	classify into	a wide range of	transform into

1. These bodies cannot be considered truly incorrupt since the tissues are _______ another substance.
2. There is now widespread support for these proposals, ___________ a recent public opinion poll.
3. One of the first commercial products to ___________ this biotechnology is likely to be genetically engineered tomatoes.
4. Acting is emerally ___________ the two types of singing and acting, and there are plays with emphasis on singing, and other dramas with emphasis on acting.
5. Employers usually decide within five minutes whether someone__________ the job.
6. The numbers of damaging insect species ______________ pesticide have multiplied from 160 to 450 since 1960.

7. AI has been applied in ____________ fields to perform specific tasks, including education, finance, heavy industry, transportation, and so on.
8. The committee ____________ representatives from both the public and private sectors.

Translation

Ⅰ. Translate the following paragraph into Chinese.

For thousands of years, people have developed creative ways to produce textiles. The use of textiles links the myriad cultures of the world and defines the way they clothe themselves, adorn their surroundings and go about their lives. The first use of textiles, most likely felt, dates back to the late Stone Age in the Middle East, roughly 100,000 years ago. A textile is a piece of cloth that has been formed by weaving, knitting, pressing or knotting together individual pieces of fiber. Yarn, a general term for long pieces of interlocked fibers, can be made from natural materials such as cotton, linen, silk and wool. Or it can be made from manufactured materials such as nylon, acrylic and polyester. The paints that give colour to yarn are called dyes. Many people today might not think much about the shirt, pants, or socks they are wearing. Manufacturing cloth is now a very low-cost process. But this was not always the case. Until the 19th century, all cloth was made by hand. It took a great deal of time and effort to gather fibers from plants or animals to make into yarn which could then be made into cloth. Humans probably first made textiles to meet important needs, including textiles for keeping warm, creating shelter, and holding goods. But cultures around the world also developed methods of making cloth that were artistic, creative, and beautiful.

Ⅱ. Translate the following paragraph into English.

一想到中国，人们脑海中首先闪现的便是丝绸。中国是丝绸的发源地，探索丝绸奥秘最好的地方是苏州。正是在那里，诞生了第一批丝绣 (silk embroidery)，如今苏州仍在生产中国最好的丝绸。丝绸可以用合成纤维和人造纤维等制成，但最优质的丝绸是由桑蚕丝 (mulberry silk) 制成的。苏州丝绸可以制成很多产品，最常见的是衣服和围巾，但也能做成鞋、餐巾、玩具，甚至风筝等。

Paragraph Writing

How to Structure a Paragraph

A paragraph usually consists of three elements: a topic sentence, supporting sentences, and a concluding sentence.

1. In the **topic sentence** (which is often the introductory sentence), the topic or focus of the paragraph is presented. It is the most general sentence of the paragraph that gives the writer's main idea or opinion about the topic and helps the reader understand what the paragraph is going to talk about. The topic sentence serves as a focal point, foregrounding the content of the whole paragraph. Although it is possible for the topic sentence to appear anywhere in a paragraph, it usually appears at the beginning.

A good topic sentence does the following:

- indicates what is in the rest of the paragraph
- contains both a topic and an opinion
- is clear and easy to follow
- does not include supporting details
- engages the reader by using interesting vocabulary

For example:

(1) Development of the Alaska oil fields created many problems for already-endangered wildlife.

This sentence introduces the topic and the writer's opinion. After reading this sentence, a reader might reasonably expect the writer to go on to provide supporting details and facts, such as what the problems are and how they were created. The sentence is clear and the word choice is interesting.

(2) I think that people should not take their pets to work, even for special occasions, because it is disruptive and someone might get bitten by a dog or a rabbit.

Even though the topic and opinion are evident, there are too many details (under what conditions, types of pets, different consequences). The phrase "special occasions" is vague. "I think" is usually unnecessary in academic writing. Therefore, the topic sentence might be revised like this: *People should not take their pets to work.*

2. The main part of the paragraph consists of **supporting sentences**: this is where the argument that explains and/or proves the topic sentence is delivered. These are sentences that convey more detailed ideas that follow the topic sentence. For example, a paragraph on the topic of people continuing to work into their 70s might have a topic sentence like this:
Retirement is a moving target for many older Americans.

Supporting sentences could include any of the following:
Fact: *Many families now rely on older relatives for financial support.*
Reason: *The life expectancy for an average American is continuing to increase.*
Statistic: *More than 20 percent of adults over age 65 are currently working or looking for work in the United States.*
Quotation: *"Stabilizing Social Security will help seniors enjoy a well-deserved retirement," said Senator Kennedy.*
Example: *Last year, my grandpa took a job with Walmart.*

A topic sentence guides the reader by signposting what the paragraph is about. All the rest of the paragraph should relate to the topic sentence.

3. The **concluding sentence** may be found as the last sentence of a paragraph, which draws together the ideas raised in the paragraph, reminds readers of the main point without repeating the same words or just gives a final comment about the topic. Don't introduce new ideas in a conclusion, which will just confuse the reader.

Concluding sentences can do any of the following:
- summarise the key points in the paragraph
- draw a conclusion based on the information in the paragraph
- make a prediction, suggestion, or recommendation about the information.

Write a paragraph on one of the following topics. One topic has an outline that you can follow.

Topic: How to deal with stress in college **Topic sentence:** There are some good ways to relieve stress for college students. **Three ways to relieve stress:** • doing regular exercise • talking with close friends. • receiving emotional therapies **Concluding sentence:** As long as you cope with stress properly, you can turn the pressure you encounter into motivation.	

More topics:

- How to choose a suitable university
- How to best handle the relationship between parents and children

Hemp Fiber

1 Natural and **organic** fibers become more and more popular these years. Most people come to realize that nature, soft and health are the most

important things of the textile. Hemp fiber is naturally one of the most environmentally friendly fibers and also the oldest.

2 Derived from the hemp plant Cannabis sativa, hemp is one of the bast fibers known from ancient times for end uses from textiles, ropes and sail cloth, to **matrices** for industrial products in the modern age, from eco-fashion **apparel** to household **décor**. In their raw state, hemp fibers are yellowish grey to deep brown.

3 Prior to Levis Strauss' **ingenious** use of hemp to create his first jean, hemp **had been** largely used as an industrial fiber, but soon became popular in the textile world. Nowadays, hemp fibers, recognized as a sustainable fabric that is **exceptionally** strong, have received wide acceptance as reinforcements in composite materials on account of their biodegradability and low **density**, which also possess **inherent** mechanical, thermal, and **acoustic** properties.

4 As the **premier** plant fiber, the **significance** of hemp to the economic and day-to-day lives of our ancestors is increasingly being recognized. Materials made from hemp have been discovered in tombs dating back to 8000 BCE. Hemp was primarily used in making sails and ropes for ships. In fact, the ships on which Christopher Columbus sailed to America in the 15th century were rigged with hemp. Indeed, hemp was so important in England in the 16th century that King Henry VIII passed an act on **parliament** which fined farmers who failed to grow the crop. In Europe, hemp fibers have been used in **prototype** quantities to **strengthen** concrete, and in other composite materials for many construction and manufacturing applications.

5 Besides fabrics, hemp is also used in the production of paper. The first identified coarse paper, made from hemp, dates to the early Western Han Dynasty. Until 1883, between 75% and 90% of all paper in the world was made with hemp fiber. Compared with wood-based paper, which led to mass **deforestation** on a large scale, hemp with its high **recyclability**

would serve as an organic and natural resource of paper, one of the most indispensable materials of everyday life.

6 For thousands of years hemp was traditionally used as an industrial fiber. Sailors relied upon hemp **cordage** for strength to hold their ships and sails, and the **coarseness** of the fiber made hemp useful for canvas, sailcloth, sacks, rope, and paper. While hemp fiber was the first choice for industry, the coarseness of the fiber restricted hemp from apparel and most home uses. Hemp needed to be softened. Traditional methods to soften vegetable fibers used **acids** to remove lignin, a type of natural glue found in many plant fibers. While this method to remove lignin worked well with cotton or flax, it **weakened** the fibers of hemp and left them too unstable for use. Hemp therefore remained as an industrial fabric.

7 In the mid 1980s, researchers developed an **enzymatic** process to successfully remove lignin from the hemp fiber without **compromising** its strength. For the first time in history, **degummed** hemp fiber could be spun alone or with other fibers to produce textiles for apparel. This technological **breakthrough** has **catapulted** hemp to the forefront of modern textile design and fashion. Given hemp's **superiority** to other fibers, the benefits of this breakthrough are enormous.

8 Hemp is a superior fiber known for its strength and **durability**. Products made from hemp will **outlast** their competition by many years. Not only is hemp strong, but it also holds its shape, stretching less than any other natural fiber. This prevents hemp garments from stretching out or becoming **distorted** with use. Due to its **porous** nature, hemp is more water absorbent, which allows it to "breathe," so that air trapped in the fibers is warmed by the body, making hemp garments naturally warm in cooler weather. One particularly unique advantage of this fiber is that it is effectively resistant to **ultraviolet** light, a major cause of cancer.

9 Hemp fiber is also **hypoallergenic** and hence suitable for **sensitive** people; it dyes and **retains** colour better than any fabric including cotton. Hemp is

an extremely fast-growing crop, producing 250% more fiber than cotton and 600% more fiber than flax using the same amount of land. The amount of land needed for obtaining equal yields of fiber places hemp at an advantage over other fibers.

10 Obtained organically and more **ecologically** sustainable than other chemically synthesised fibers, hemp is extremely **versatile** and can be used for countless products such as apparel, accessories, shoes, furniture, and home furnishings. As a fabric, hemp provides all the warmth and softness of a natural textile but with a superior durability seldom found in other materials. Apparel made from hemp **incorporates** all the **beneficial** qualities and will likely last longer and **withstand** harsh conditions. Hemp blended with other fibers easily incorporates the **desirable** qualities of both textiles. The soft elasticity of cotton or the smooth texture of silk combined with the natural strength of hemp creates a whole new genre of fashion design, such as hemp silk Charmeuse. A naturally high luster of hemp helps to make hemp fabrics truly shine.

11 For various properties, hemp is definitely a fabric with excellent scope of sustainability and **immense** possibilities. It is likely that hemp will eventually **supersede** cotton, linen, and polyester in numerous areas. As the modern world seeks new organic and sustainable materials for everyday use, this versatile crop has immense **potential** to be elevated as the ideal fashion choice for **interiors** and clothing. In a nutshell, hemp textiles are the wave of the future and a way to a good life!

Notes

Cannabis sativa 大麻

Christopher Columbus 克里斯多弗·哥伦布（1451—1506，意大利航海家）

King Henry VIII 亨利八世国王（1491—1547）

Charmeuse 查米尤斯绉缎，一种具有缎面效果的轻质织物

New words and phrases

organic /ɔːˈɡænɪk/ — *a.* ① produced by or from living things 有机物的；生物的 ② (of food, farming methods, etc.) produced or practiced without using artificial chemicals 有机的；不使用化肥的；绿色的

matrix /ˈmeɪtrɪsiːz/ — *n.* [C] (*pl.* matrices) a mould in which sth is shaped 铸模；模子；模具

apparel /əˈpærəl/ — *n.* [U] ① clothing, when it is being sold in shops/stores （商店出售的）衣服，服装② (*formal*) clothes, particularly those worn on a formal occasion（尤指正式场合穿的）衣服，服装

décor /ˈdeɪkɔː(r)/ — *n.* [C, usually sing.] the style in which the inside of a building is decorated（建筑内部的）装饰布局，装饰风格

ingenious /ɪnˈdʒiːniəs/ — *a.* very clever and involves new ideas, methods, or equipment 灵巧的；新颖的

exceptionally /ɪkˈsepʃənəli/ — *ad.* used before an adjective or adverb to emphasise how strong or unusual the quality is（用于形容词和副词之前表示强调）罕见；特别，非常

density /ˈdensəti/ — *n.* ① [U] the quality of being dense; the degree to which sth is dense 密集；稠密；密度；浓度 ② [C, U] (*physics*) the thickness of a solid, liquid or gas measured by its mass per unit of volume 密度（固体、液体或气体单位体积的质量）

inherent /ɪnˈhɪərənt/ — *a.* ~ (**in sb/sth**) that is a basic or permanent part of sb/sth and that cannot be removed 固有的；内在的

acoustic /əˈku:stɪk/ — *a.* ① of or relating to the scientific study of sound 声学的 ② related to sound or to the sense of hearing 声音的；音响的；听觉的

premier /ˈpremiə(r)/ — *a.* [only before noun] most important, famous or successful 首要的；最著名的；最成功的；第一的

n. [C] the leader of the government of a country 首相，总理

significance /sɪgˈnɪfɪkəns/ — *n.* [C, U] the importance of sth, especially when this has an effect on what happens in the future（尤指对将来有影响的）重要性，意义

parliament /ˈpɑ:ləmənt/ — *n.* [C, sing.+sing./pl. v.] the group of people who are elected to make and change the laws of a country 议会；国会

prototype /ˈprəʊtətaɪp/ — *n.* [C] ~ (**for/of sth**) the first design of sth from which other forms are copied or developed 原型；雏形；最初形态

strengthen /ˈstreŋθən/ — *v.* to become stronger; to make sb/sth stronger 加强；增强；巩固

deforestation /ˌdi:ˌfɒrɪˈsteɪʃn/ — *n.* [U] the act of cutting down or burning the trees in an area 毁林；滥伐森林；烧林

recyclability /ri:sɪklə'bɪlɪtɪ/ — *n.* [U] the ability to be recycled 再循环能力

cordage /'kɔ:dɪdʒ/ — *n.* [C] the lines and rigging of a vessel 绳索；纤维绳

coarseness /ˈkɔːsnəs/ *n.* [U] looseness or roughness in texture (as of cloth) 粗；粗糙

acid /ˈæsɪd/ *n.* [C, U] a chemical substance that has a pH of less than seven 酸

weaken /ˈwiːkən/ *v.* to make sb/sth less strong or powerful; to become less strong or powerful （使）虚弱，衰弱；减弱；削弱

enzymatic /ˌenzaɪˈmætɪk/ *a.* of, relating to, or produced by an enzyme 酶的

compromise /ˈkɒmprəmaɪz/ *v.* ① risk harming or losing sth important 危及；损害 ② ~ **(with sb) (on sth)** to give up some of your demands after a disagreement with sb, in order to reach an agreement（为达成协议而）妥协，折中，让步 ③ ~ **(on sth)** to do sth that is against your principles or does not reach standards that you have set 违背（原则）；达不到（标准）

n. [C] an agreement made between two people or groups in which each side gives up some of the things they want so that both sides are happy at the end 妥协；折中；互让；和解

degum /diːˈgʌm/ *v.* to remove or get rid of gum 使脱胶；使去胶

breakthrough /ˈbreɪkθruː/ *n.* [C] an important development that may lead to an agreement or achievement 重大进展；突破

catapult /ˈkætəpʌlt/

v. ① If something catapults you into a particular state or situation, or if you catapult there, you are suddenly and unexpectedly caused to be in that state or situation. 使突然处于；突然处于 ② to throw sb/sth or be thrown suddenly and violently through the air （被）猛掷，猛扔

superiority /sjuːˌpɪəriˈɒrəti/

n. [U] ① ~ **(in sth)** | ~ **(to/over sth/sb)** the state or quality of being better, more skillful, more powerful, greater, etc. than others 优越（性）；优势 ② behaviour that shows that you think you are better than other people 优越感；神气活现的样子；盛气凌人的行为

durability /ˌdjʊərəˈbɪləti/

n. [U] the ability of staying in good condition for a long time, even if used a lot 耐久性；坚固

distort /dɪˈstɔːt/

v. ① to change the shape, appearance or sound of sth so that it is strange or not clear 使变形；扭曲；使失真 ② to twist or change facts, ideas, etc. so that they are no longer correct or true 歪曲；曲解

porous /ˈpɔːrəs/

a. having many small holes that allow water or air to pass through slowly 多孔的；透水的；透气的

ultraviolet /ˌʌltrəˈvaɪələt/

a. [usually before noun] (*abbr.* **UV**) of or using electromagnetic waves that are just shorter than those of violet light in the spectrum and that cannot be seen 紫外线的；利用紫外线的

hypoallergenic /ˌhaɪpəʊˌæləˈdʒenɪk/

a. (of cosmetics, earrings, etc.) not likely to cause an allergic reaction (化妆品、耳环等）不会导致过敏反应的

sensitive /ˈsensətɪv/ *a.* ① ~ **(to sth)** reacting quickly or more than usual to sth 敏感的；过敏的 ② ~ **(to sth)** aware of and able to understand other people and their feelings 体贴的；体恤的；善解人意的

retain /rɪˈteɪn/ *v.* ① to keep sth; to continue to have sth 保持；持有；保留；继续拥有 ② to continue to hold or contain sth 保持；继续容纳

ecologically /ˌiːkəˈlɒdʒɪkli/ *ad.* with respect to ecology 从生态学的观点看

versatile /ˈvɜːsətaɪl/ *a.* ① (of food, a building, etc.) having many different uses 多用途的；多功能的 ② (of a person) able to do many different things 多才多艺的；有多种技能的；多面手的

incorporate /ɪnˈkɔːpəreɪt/ *v.* ~ **sth (in/into/within sth)** to include sth so that it forms a part of sth 将…包括在内；包含；吸收；使并入

beneficial /ˌbenɪˈfɪʃl/ *a.* ~ **(to sth/sb)** improving a situation; having a helpful or useful effect 有利的；有裨益的；有用的

withstand /wɪðˈstænd/ *v.* (**withstood, withstood**) (*formal*) to be strong enough not to be hurt or damaged by extreme conditions, the use of force, etc. 承受；抵住；顶住；经受住

desirable /dɪˈzaɪərəbl/ *a.* ~ **(that) ...** | ~ **(for sb) (to do sth)** (*formal*) that you would like to have or do; worth having or doing 向往的；可取的；值得拥有的；值得做的

immense /ɪˈmens/ *a.* extremely large or great 极大的；巨大的

supersede /ˌsuːpəˈsiːd/	*v.* [often passive] to take the place of sth/sb that is considered to be old-fashioned or no longer the best available 取代，替代（已非最佳选择或已过时的事物）
potential /pəˈtenʃl/	*n.* [U] ① ~ **(for/for doing sth)** the possibility of sth happening or being developed or used 可能性；潜在性 ② qualities that exist and can be developed 潜力；潜质 *a.* [only before noun] that can develop into sth or be developed in the future 潜在的；可能的
interior /ɪnˈtɪərɪə/	*n.* [C, usually sing.] the inside part of sth 内部；内饰；室内设计
from … to …	used to mention the two ends of a range（表示幅度或范围）从……到
prior to	in advance of; before 居先；在……之前
on account of	because of something else, especially a problem or difficulty 由于；鉴于；因为
date back to	to have existed since a particular time in the past 追溯到
restrict sb/sth from doing sth	to stop sb/sth from doing sth freely 束缚；妨碍；阻碍
remove from	to get rid of sth so that it does not exist any longer 去除；剔除；使消失
at/in/to the forefront of sth	in or into an important or leading position in a particular group or activity 处于最前列；进入重要地位（或主要地位）
at an advantage over	a favourable or superior position when compared someone or something else 比……有优势

combine ... with to add or mix two or more things together 将……结合

in a nutshell (to say or express sth) in a very clear way, using few words 简而言之

Reading Comprehension

Understanding the text

Choose the best answer to each of the following questions.

1. Which of the following statements about hemp fiber is true?
   A. Hemp fiber is the most environmentally friendly man-made fiber and also the earliest.
   B. Hemp fiber can only be used for the production of textiles, ropes and sail cloth, etc.
   C. Hemp fibers are widely used as reinforcements in composite materials for biodegradability and low density.
   D. Hemp fibers do not possess mechanical, thermal, and acoustic properties.
2. Which of the following cannot be used as evidence to show the significance and ancient history of hemp fiber?
   A. Christopher Columbus sailed to America on ships rigged with hemp in the 15th century.
   B. Materials made from hemp have been discovered in tombs dating back to 8000 BCE.
   C. The first identified coarse paper, made from hemp, dates to the early Western Han Dynasty.
   D. Levis Strauss ingeniously used hemp to create his first jean.
3. Which of the following is not the reason why hemp was long limited for industrial uses?
   A. The coarseness of hemp made it useful for canvas, sailcloth, sacks, rope, and paper.
   B. Traditional methods to soften vegetable fibers are not suitable for hemp.

C. Researchers developed an enzymatic process to successfully remove lignin from the hemp fiber.

D. The coarseness of the fiber restricted it from apparel and most home uses.

4. Which of the following is not the property of hemp?

A. Durability.

B. Sensitivity.

C. Breathability.

D. Sustainability.

5. According to the last two paragraphs, which of the following statements is correct?

A. Hemp possesses a superior durability, but cannot provide warmth and softness.

B. Apparel made from hemp will surely last longer and withstand harsh conditions.

C. Hemp blended with other fibers such as cotton or silk can incorporate the favourable qualities of both.

D. For various properties, hemp will definitely supersede cotton, linen, and polyester in numerous areas.

Critical thinking

Work in pairs and complete the following tasks.

1. Hemp is one of the oldest and most important plant fibers. Please give a brief comment on the properties and uses of hemp fiber.
2. There are various kinds of textile fibers with unique characteristics. What do you think are the factors to be considered for a preferable textile fiber?

Research project

To keep up with the growth and focus on more environment-friendly fibers, textile companies are developing fibers out of recycled materials and natural sources for technical textiles. Search for information about one type of newly developed sustainable fibers and then summarise your findings in a research report and present it in class.

Language Enhancement

Words in use

Fill in the blanks with the words given below. Change the form when necessary. Each word can be used only once.

significance	inherent	ingenious	breakthrough	superiority
strengthen	organic	compromise	beneficial	withstand

1. It is reported that more and more people prefer to buy ________ food even though it is more expensive.
2. The catalogue is full of ________ ideas for transforming your house into a dream home.
3. The privileges and freedoms ________ in self-government are balanced by the duties and responsibilities of citizenship.
4. Since education is of great ________, the plan of reforming the current education system has been under discussion for a year.
5. Any experience can teach and ________ you, but particularly the more difficult ones.
6. The employers will have to be ready to ________ if they want to avoid a strike.
7. Scientists have made a major ________ in the treatment of cancer, bringing great hopes to those patients.
8. The sense of ________ I gained was essentially related to control and, eventually, to the triumph of the will.
9. It was challenging for the locals to build a bridge that was strong enough to ________ the current of the Yellow River.
10. While a moderate amount of stress can be ________, too much stress can exhaust you.

Expressions in use

Fill in blanks with the expressions given below. Change the form where necessary. Each expression can be used only once.

prior to	on account of	restrict ... from	in a nutshell
remove from	date back to	at the forefront of	combine with

1. Sleeping pills can interfere with the effect of other medications and can be dangerous when ____________ them.
2. It was necessary that this should be done in one single day ____________ the extreme urgency and rigor of events.
3. The relics were ____________ the house and taken to a local museum for identification.
4. If a person places money ____________ anything else, it's very likely that he could not enjoy the happiness in life.
5. When an employing unit recruits female workers, it shall not have such provisions as ____________ female workers ____________ getting married or bearing a child included in the labor contract.
6. The sale also features Britain's most extensive collection of historical cookbooks, some of which ____________ the reign of Henry VIII.
7. A study of women at work says, ____________, that opportunities have opened up dramatically.
8. Thomas Edison was ____________ the search for alternative ways to power vehicles, a search that continues today.

Sentence structure

I. Complete the following sentences by translating the Chinese sentences into English, using "given +n./that ... " structure.

Model: ________________________（鉴于麻纤维相对于其他纤维的优势）, the benefits of this breakthrough are enormous.

→ Given hemp's superiority to other fibers, the benefits of this breakthrough are enormous.

1. ______________________________（考虑到她喜欢孩子）, teaching seems to be a suitable job for her.
2. ______________________________（鉴于公众对生态学方面的问题非常敏感）, the investment in copper mines, forestry, or land may evoke local fears.
3. ______________________________（考虑到病人有一些残疾）, we still try to enable them to be as independent as possible.

II. Rewrite the following sentences by using "It's (quite/very/highly/extremely/ more/most/less/least/hardly) likely that ... " structure.

Model: Hemp will eventually be likely to supersede cotton, linen, and polyester in numerous areas.

→ It is likely that hemp will eventually supersede cotton, linen, and polyester in numerous areas.

1. The jury is very likely to believe he was in the apartment at the time of the crime.

____

____.

2. These changes are highly likely to impoverish single-parent families even further.

____

____.

3. If the stress outweighs the benefits, your goal is not likely to be a healthy one.

____

____.

Modal Fabric

1 In recent years, modal has become so popular in the world that it is very likely that you have several pieces of clothing or linen made of modal fabric. But what is really modal and why is it quite so popular?

2 Modal is a type of rayon, a semi-synthetic cellulose fiber made by spinning reconstituted cellulose from beech trees. The wood fibers are pulped into liquid form and then forced through tiny holes, creating the fiber. This is then woven together to make the modal fabric, which may be used on its own or in a textile blended with other materials like cotton or spandex in household items such as pajamas, towels, bathrobes, underwear and bedsheets. Modal is the generic name for a semi-synthetic upgrade to viscose rayon that eliminates some of the most wasteful or harmful aspects of the viscose production process. Most consumers and manufacturers also agree that modal rayon is a structurally superior product to viscose rayon. Essentially modal is processed under different conditions to produce a fiber that is stronger and more stable than traditional rayon, yet has a soft feel, similar to cotton.

3 Modal was originally developed in 1951 in Japan, but today most of the modal is produced by an Austrian company known as Lenzing AG who has been the primary producer of modal fabrics since 1964. This textile giant is based in Europe, but it has factories all over the world, mainly in China. Other countries where this fabric is produced include India, Pakistan, Indonesia, Germany and UK.

4 Even though modal has been around for so many decades, it is now more popular than ever and more modal is produced. There are, indeed, many advantages to using modal as a fabric. For one thing, lightweight,

breathable, thin and with an impressive stretch and softness, modal rayon is highly popular in sportswear. For instance, it's common to see this fabric used for yoga pants, bike shorts, and even swimwear. Modal rayon wicks sweat effectively and is easy to clean, surprisingly durable and long-lasting. For another, modal is about fifty percent more absorbent to water than cotton, which makes it ideal for activewear.

5 Due to its silky texture, it is a common choice for other forms of household textiles, such as bed sheets, that are kept close to the skin for prolonged periods of time. Another reason why this fabric is so popular these days is that it resists shrinkage and pilling a lot better than many of its counterparts. Like all other cellulose fibers, modal is designed both to stay colourfast and absorb the dye when washed in warm water, making it a popular choice for underwear and activewear alike.

6 Modal is categorised as a semi-synthetic fiber rather than a natural one because of the involvement of a number of chemicals during the manufacturing process. Is there an environmental impact in the production of this fabric? All things considered, modal is not inherently sustainable or environmentally friendly, but it has the potential to be under certain circumstances. According to the Sewport guide on modal, "It is up to individual manufacturers to follow the manufacturing processes that will result in environmentally-friendly fabrics." So, there are two main environmental issues to be aware of when it comes to modal. As modal is created from tree fibers, consumers should be aware of both the source of the trees, and the effects of processing them into fiber.

7 When it comes to Lenzing Modal, the Austrian company claims to have a positive environmental footprint with carbon neutral. Their process ensures smaller carbon footprints without compromising the quality of their end product. Besides, Lenzing came up with a more environment-friendly method to bleach the wood pulp. Being aware of the environmental impact and modifying it to ensure the long-lasting health of our planet have certainly brought them a lot of reliability. That is why now Lenzing Modal

is a signature fabric with impressive customer demand around the globe.

8 However, a good number of disreputable manufacturers may cut corners in the production process to keep costs low, but doing so may result in a product that is poor in quality or harmful to the environment. Deforestation is a major force behind their mass productions. On the other hand, modal is said to be slightly more eco-friendly than viscose and cotton in that the production of modal requires considerably less land per ton, consumes between ten to twenty times less water and uses less chemicals.

9 Above all, many rayon manufacturers have started to use chemical scrubbers or machines to trap the chemicals before they make their way into the ecosystem, further reducing harm. It is important to consider both the source materials and fiber processing, as well as the weaving, cutting, and transportation process when thinking about the sustainability of a fabric.

Unit 3

Textile Craft and Technology

Every time that I wanted to give up, if I saw an interesting textile, print what ever, suddenly I would see a collection.

—*Anna Sui*

Light and colour are closely linked. The colours can make a crucial change in nature, if you switch from daylight to artificial light or just from strong to weak illumination. In addition, colour perception is affected by the material structure. Even if a piece of textile can have the same colour as a shiny enamel plate, then they will act completely different.

—*Verner Panton*

Pre-Reading Activities

1. Watch the video and choose the best answer to each of the following questions.

(1) What are the two widely used dyes of batik?

A. Bee's wax and indigo.

B. Acid dyes and indigo.

C. Bee's wax and acid dyes.

(2) Which of the following is the correct process of producing batik?

A. Washing the cloth, spreading wax, painting, and dyeing.

B. Painting, spreading wax, dyeing and washing the cloth.

C. Spreading wax, painting, dyeing and washing the cloth.

2. Watch the video again and fill in the blanks based on what you hear.

Batik (la ran) is a (1) ________ painting and textile dyeing (2) ________ with a history of more than 2,000 years. Bee's wax and indigo are two widely used dyes. A steel knife is used to draw (3) ________ on the cloth. A piece of la ran (batik) is (4) ________ after spreading wax, (5) ________, dyeing and (6) ________ the cloth. The most interesting part is that (7) ________ grains form along the cracks of the cloth after cooling. Different (8) ________ come from the same patterns: flowers, birds, fish, (9) ________, and geometrical patterns on the cloth look (10) ________ and unsophisticated.

3. Discuss the following questions with your partner.

(1) Have you tried any textile crafts and technologies such as batik or embroidery? If so, please share your experience with your partner.

(2) With the development of technology, do you still believe that traditional textile crafts are necessary in our daily life? Why or why not?

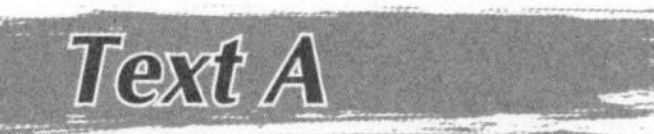

Embroidery—An Art of Expression

1 Clothing was often used as a medium of expression with its own form and **techniques**. The lines, **textures**, colours, **proportions** and scale of fabric designs and the shapes of garments can and have varied enormously at different times and different places throughout history. Ideas of human beauty change with changes in the **Zeitgeist**. Often clothing is used by individuals as part of an **attempt** to conform to the physical ideal of human body at a particular place.

2 As the time passed, different types of fabrics were discovered. Varied styles of garment **silhouettes** were developed. Everyone used their creative ability in decorating their **attire**. This creative **enthusiasm** of the people gave way for unending surface enrichment-techniques. Different forms of weaving, dyeing, printing, knitting, **crocheting**, **macramé**, and embroideries were **employed** to design various styles.

3 Embroidery is the art of decorating a ground fabric with stitches to enrich it and add to its beauty. It can be worked on any **pliable** material from leather to linen, in threads ranging from **wool** to the finest gold and it can be embellished with jewels, pearls and **enamels**. In this way, it has been practiced for **decades**.

4 The origin of the embroidery, as of all needlework, do **obscure** owe to **perishable** nature of fabric. It is difficult to give an exact history of it. The origin of embroidery can be dated back to Cro-Magnon days or 30000 BCE. During a recent archaeological find, **fossilised** remains of heavily hand-stitched and decorated clothing, boots and a hat were found.

5 In Siberia, around 5000 BCE and 6000 BCE elaborately drilled shells stitched with decorative designs onto animal hides were discovered.

Chinese thread embroidery dates back to 3500 BCE where pictures **depict** embroidery of clothing with silk thread, **precious** stones and pearls. Examples of surviving Chinese chain stitch embroidery worked in silk thread have also been found and dated to the Warring States period (5th–3rd century BCE).

6 Embroidery and most other fibre and needlework arts are believed to **originate** in the Orient and Middle East. **Primitive** humankind quickly found that the stitches used to join animal skins together could also be used for embellishment. Recorded history, **sculptures**, paintings and vases depicting **inhabitants** of various ancient civilisations show people wearing thread-embroidered clothing.

7 During the 1100s, smaller seed pearls were sewn on **vellum** to decorate religious items and from the 1200s through 1300s beads were **embroidered** onto clothing. By 1500 CE, embroideries had become more **lavish** in Europe, as well as other areas of the world. From this period through the 1700s elaborate thread and **bead** embroidery gained popularity. Bead embroidery could be found on **layette** baskets, court dress, home furnishings and many other items.

8 Elaborately **embroidered** clothing, religious objects, and household items have been a mark of wealth and status in many cultures including ancient Persia, India, China, Japan, Byzantium, and **medieval** and Baroque Europe. Traditional folk techniques were passed from generation to generation in cultures as diverse as northern Vietnam, Mexico, and eastern Europe. Professional workshops and **guilds** arose in medieval England. The output of these workshops, called Opus Anglicanum or "English work," was famous throughout Europe. The manufacture of machine-made embroideries in St. Gallen in eastern Switzerland **flourished** in the latter half of the 19th century.

9 The process used to tailor, patch, mend and reinforce cloth later **fostered** the development of sewing techniques, and the decorative possibilities of

sewing led to the art of embroidery. Elaborate **freehand** stitched thread embroidery began to **dwindle** with the machine age of the 1800s when Art needlework and Berlin wool-work appeared on the scene. Berlin wool-work, **canvas** thread embroidery, was popular through the 1870s only to be replaced in popularity by counted cross-stitch of the 1880s, using square **meshed** canvas with stitch-by-stitch thread designs. With the introduction of printed patterns in colour, the need for counting each stitch was pass in many instances. Although elaborate freehand thread embroidery was **waning** in popularity, bead embroidery was beginning its **heyday** along with the new needlework stitches of the 1800s.

10 The fabrics and yarns used in traditional embroidery vary from place to place. Wool, linen, and silk have been in use for thousands of years for both fabric and yarn. Today, embroidery thread is manufactured in cotton, **rayon**, and novelty yarns as well as in traditional wool, linen, and silk. **Ribbon** embroidery uses narrow ribbon in silk or silk **organza** blend ribbon, most commonly to create floral **motifs**.

11 Surface embroidery techniques such as chain stitch and couching or laid-work are the most **economical** of expensive yarns; couching is generally used for gold work. Canvas work techniques, in which large amounts of yarn are buried on the back of the work, use more materials but provide a **sturdier** and more **substantial** finished textile.

12 Much contemporary embroidery work is stitched with a computerised embroidery machine using patterns "digitised" with embroidery software. In machine embroidery, different types of "fills" add texture and design to the finished work. Machine embroidery is used to add logos and **monograms** to business shirts or jackets, gifts, and team apparel as well as to decorate household linens, **draperies**, and decorator fabrics that **mimic** the elaborate hand embroidery of the past. Many people are choosing embroidered logos placed on shirts and jackets to promote their company. Yes, embroidery has come a long way, both in style, technique and use. It also appears to maintain its **intrigue** as its popularity continues to grow with it.

13 Embroidery embraces a wealth of different historical and contemporary techniques. All over the world people continue to develop traditional and **innovative** methods to decorate garments and soft furnishings, or to create an art form that reinterpretes the traditional skills, thus ensuring that embroidery is a constantly growing and developing craft.

Notes

Cro-Magnon Cro-Magnon refers to a hominid of a tall erect race of the Upper Paleolithic Period (c. 40,000 to c. 10,000 years ago) known from skeletal remains found chiefly in southern France and is classified as the same species (Homo sapiens) as present-day humans.

The Warring States period The period of the Warring States (Zhanguo or Chan-Kuo) refers to the era of about 475 BCE to 221 BCE. It commenced at a time when the numerous petty city-state kingdoms of the Spring and Autumn period had been consolidated into seven major contenders and a few minor enclaves.

Siberia 西伯利亚

Persia 波斯（伊朗在欧洲的古希腊语和拉丁语的旧称译音）

Byzantium 拜占庭（古罗马城市，今称伊斯坦布尔）

Opus Anglicanum（产于英格兰的）优质教堂礼服，多有使用贵重材料（特别是银线）

New words and phrases

technique /tekˈniːk/ *n.* ① [C] a particular way of doing sth, especially one in which you have to learn special skills 技巧；技艺；工艺 ② [U. sing.] the skill with which sb is able to do sth practical 技术；技能

texture /ˈtekstʃə(r)/ *n*. [C, U] the way a surface, substance or piece of cloth feels when you touch it, for example how rough, smooth, hard or soft it is 质地；手感

proportion /prəˈpɔːʃn/ *n*. ① [U] [C, usually pl.] the correct relationship in size, degree, importance, etc. between one thing and another or between the parts of a whole 正确的比例；均衡；匀称 ② [pl.] the measurements of sth; its size and shape 面积；体积；规模；程度

Zeitgeist /ˈzaɪtgaɪst/ *n*. [sing.] (*from German, formal*) the general mood or quality of a particular period of history, as shown by the ideas, beliefs, etc. common at the time 时代精神；时代思潮

attempt /əˈtempt/ *n*. [C, U] ~ **(to do sth)** | ~ **(at sth/at doing sth)** an act of trying to do sth, especially sth difficult, often with no success 企图；试图；尝试

v. to make an effort or try to do sth, especially sth difficult 努力；尝试；试图

silhouette /ˌsɪluˈet/ *n*. [C] the shape of a person's body or of an object（人的）体形；（事物的）形状

attire /əˈtaɪə(r)/ *n*. [U] (*formal*) clothes, especially fine or formal ones（尤指华丽或正式的）服装；衣服

enthusiasm /ɪnˈθjuːziæzəm/ *n*. [U] ~ **(for sth/for doing sth)** a strong feeling of excitement and interest in sth and a desire to become involved in it 热情；热心；热忱

crochet /ˈkrəʊʃeɪ/ *n*. [U] a way of making clothes, etc. from wool or cotton using a special thick needle with a hook at the end to make a pattern of connected threads 钩针编织

v. to make sth using crochet 用钩针编织

macramé /məˈkrɑːmi/ *n.* [U] the art of tying knots in string in a decorative way, to make things 装饰编结艺术；编结艺术

employ /ɪmˈplɔɪ/ *v.* (*formal*) to use sth such as a skill, method, etc. for a particular purpose 应用；运用；使用

pliable /ˈplaɪəbl/ *a.* easy to bend without breaking 易弯曲的；柔韧的

wool /wʊl/ *n.* [U] ① the soft fine hair that covers the body of sheep, goats and some other animals（羊等动物的）绒，毛 ② long thick thread made from animal's wool, used for knitting 毛线；绒线

enamel /ɪˈnæml/ *n.* [U, C] a substance that is melted onto metal, pots, etc. and forms a hard shiny surface to protect or decorate them; an object made from enamel 搪瓷；珐琅；搪瓷制品

decade /ˈdekeɪd/ *n.* [C] a period of ten years, especially a period such as 1910–1919 or 1990–1999 十年，十年期

obscure /əbˈskjʊə(r)/ *a.* not well known 无名的；鲜为人知的

perishable /ˈperɪʃəbl/ *a.* (especially of food) likely to decay or go bad quickly 易腐烂的；易变质的

fossilised /ˈfɒsəlaɪzd/ *a.* ① to become hard and form fossils, instead of decaying completely 变成化石的，石化的 ② set in a rigidly conventional pattern of behaviour, habits, or beliefs 僵化的

depict /dɪˈpɪkt/ *v.* ① to show an image of sb/sth in a picture 描绘；描画 ② to describe sth in words, or give an impression of sth in words or with a picture 描写；描述；刻画

precious /ˈpreʃəs/ *a.* rare and worth a lot of money 珍奇的；珍稀的

originate /əˈrɪdʒɪneɪt/ *v.* ① (*formal*) to happen or appear for the first time in a particular place or situation 起源；发源；发端于 ② (*formal*) to create sth new 创立；创建；发明

primitive /ˈprɪmətɪv/ *a.* belonging to an early stage in the development of humans or animals 原始的；人类或动物发展早期的

sculpture /ˈskʌlptʃə(r)/ *n.* ① [C, U] a work of art that is a solid figure or object made by carving or shaping wood, stone, clay, metal, etc. 雕像；雕塑品；雕刻品 ② [U] the art of making sculptures 雕刻术；雕塑术

inhabitant /ɪnˈhæbɪtənt/ *n.* [C] a person or an animal that lives in a particular place（某地的）居民，栖息动物

vellum /ˈveləm/ *n.* [U] material made from the skin of a sheep, goat or calf, used for making book covers and, in the past, for writing on（书封或旧时书写用的）羊皮纸，犊皮纸

lavish /ˈlævɪʃ/ *a.* large in amount, or impressive, and usually costing a lot of money 大量的；给人印象深刻的；耗资巨大的

bead /biːd/ *n.* [C] a small piece of glass, wood, etc. with a hole through it, that can be put on a string with others of the same type and worn as jewellery, etc.（有孔的）珠子

layette /leɪˈet/ *n.* a set of clothes and other things for a new baby 新生儿的全套用品

embroider /ɪmˈbrɔɪdə(r)/ *v.* to decorate cloth with a pattern of stitches usually using coloured thread 刺绣

embroidered /ɪmˈbrɔɪdəd/ *a.* 绣花的，刺绣的

medieval /ˌmediˈiːvl/ *a.* connected with the Middle Ages (about 1000 CE to 1450 CE) 中世纪的（约公元 1000 年到 1450 年）

baroque /bəˈrɒk/ — *a.* used to describe European architecture, art and music of the 17th and early 18th centuries that has a grand and highly decorated style 巴罗克风格的（17至18世纪早期流行于欧洲，气势雄伟、装饰华丽的特色反映在建筑、绘画和音乐等艺术上）

guild /gɪld/ — *n.* [C] ① an organisation of people who do the same job or who have the same interests or aims （行业）协会 ② an association of skilled workers in the Middle Ages （中世纪的）行会，同业公会

flourish /ˈflʌrɪʃ/ — *v.* to develop quickly and be successful or common 繁荣；昌盛；兴旺

foster /ˈfɒstə(r)/ — *v.* to encourage sth to develop 促进；助长；培养；鼓励

freehand /ˈfriːhænd/ — *a.* drawn without using a ruler or other instruments 徒手画的；不用仪器画的

dwindle /ˈdwɪndl/ — *v.* to become gradually less or smaller （逐渐）减少，变小，缩小

canvas /ˈkænvəs/ — *n.* [U] a strong heavy rough material used for making tents, sails, etc. and by artists for painting on 帆布

mesh /meʃ/ — *v.* ① (*formal*) to fit together or match closely, especially in a way that works well; to make things fit together successfully（使）吻合，相配，匹配，适合 ② (of parts of a machine) to fit together as they move （机器零件）啮合

wane /weɪn/ — *v.* ① to become gradually weaker or less important 衰落；衰败；败落；减弱 ② (of the moon) to appear slightly smaller each day after being round and full （月亮）缺；亏

heyday /ˈheɪdeɪ/ *n.* [usually sing.] the time when sb/sth had most power or success, or was most popular 最为强大（或成功、繁荣）的时期

rayon /ˈreɪɒn/ *n.* [U] a fibre made from cellulose; a smooth material made from this, used for making clothes 人造丝；人造丝织品

ribbon /ˈrɪbən/ *n.* ① [U, C] a narrow strip of material, used to tie things or for decoration （用于捆绑或装饰的）带子；丝带 ② [C] sth that is long and narrow in shape 带状物；狭长的东西

organza /ɔːˈɡænzə/ *n.* [U] a type of thin stiff transparent cloth, used for making formal dresses 透明硬纱（用于制作礼服）

motif /məʊˈtiːf/ *n.* [C] a design or a pattern used as a decoration 装饰图案；装饰图形

economical /ˌiːkəˈnɒmɪkl/ *a.* ① providing good service or value in relation to the amount of time or money spent 经济的；实惠的 ② using no more of sth than is necessary 节俭的；节约的；简洁的 ③ not spending more money than necessary 精打细算的；省钱的

sturdy /ˈstɜːdi/ *a.* ① (of an object) strong and not easily damaged（物品）结实的；坚固的 ② (of people and animals, or their bodies) physically strong and healthy（人、动物或身体）强壮的；健壮的 ③ not easily influenced or changed by other people 坚决的；坚定的；顽强的

substantial /səbˈstænʃl/ *a.* ① large in amount, value or importance 大量的；价值巨大的；重大的 ② (*formal*) large and solid; strongly built 大而坚固的；结实的；牢固的

monogram /ˈmɒnəgræm/ *n*. [C] two or more letters, usually the first letters of sb's names, that are combined in a design and marked on items of clothing, etc. that they own 字母组合图案，交织字母，花押字（常由姓名首字母组成，标在自己的衣服等物品上）

drapery /ˈdreɪpəri/ *n*. [U] ① (also *pl*. draperies) cloth or clothing hanging in loose folds 垂褶布（或织物）② *old-fashioned* cloth and materials for sewing sold by a draper （布商出售的）织物，布料

mimic /ˈmɪmɪk/ *v*. ① to copy the way sb speaks, moves, behaves, etc., especially in order to make other people laugh 模仿（人的言行举止）；（尤指）做滑稽模仿 ② to look or behave like sth else （外表或行为举止）像，似

n. a person or an animal that can copy the voice, movements, etc. of others 会模仿的人（或动物）

intrigue /ɪnˈtriːg/ *v*. ① [often passive] to make sb very interested and want to know more about sth 激起……的兴趣；引发……的好奇心

② ~ **(with sb) (against sb)** (*formal*) to secretly plan with other people to harm sb 秘密策划（加害他人）；密谋

/ˈɪntriːg/ *n*. ① [U] the activity of making secret plans in order to achieve an aim, often by tricking people 密谋策划；阴谋 ② [C] a secret plan or relationship, especially one which involves sb else being tricked 密谋；秘密关系；阴谋诡计 ③ [U] the atmosphere of interest and excitement that surrounds sth secret or important 神秘气氛；引人入胜的复杂情节

innovative /ˈɪnəveɪtɪv/	*a.* (*approving*) introducing or using new ideas, ways of doing sth, etc. 引进新思想的；采用新方法的；革新的；创新的
part of sth	some but not all of a thing 部分
conform to/with sth	to obey a rule, law, etc. 遵守，遵从，服从（规则、法律等）
owe sth to sb/sth	to exist or be successful because of the help or influence of sb/sth 归因于；归功于
begin to do sth	start to do sth 开始做……
on the scene	When a person or thing appears on the scene, they come into being or become involved in something. 登场；到场
such as	for example 例如

Reading Comprehension

Understanding the text

Answer the following questions.

1. Why does the author say "clothing was often used as a medium of expression"?
2. According to the author, what kinds of factors are employed to design various styles throughout history?
3. What is the definition of embroidery in the text?
4. According to the text, when did embroidery originate?
5. When did the manufacture of machine-made embroideries flourish?
6. What led to the art of embroidery?
7. How do the fabrics and yarns used in traditional embroidery vary from place to place?
8. How is machine embroidery applied now?

Critical thinking

Work in pairs and discuss the following questions.

1. According to your understanding, why is embroidery an art of expression? Please share your reasons with examples.
2. How do you view the coexistence of traditional embroidery and machine embroidery at present?

Mind mapping

Read Paragraphs 4–9 carefully and try to find out the key information to complete the timeline of the development of embroidery.

1. ____________: the origin of embroidery

↓

around 5000 and 6000 BCE: 2. ____________

↓

3. ____________: embroidery of clothing with silk thread, precious stones and pearls in China

↓

during thc 1100s: 4. ____________

↓

5.____________: beads were embroidered onto clothing

↓

by 1500 CE: 6. ____________

↓

7. ____________: elaborate thread and bead embroidery gained popularity

Language Enhancement

Words in use

Fill in the blanks with the words given below. Change the form when necessary. Each word can be used only once.

economical	originate	intrigue	flourish	substantial
depict	dwindle	technique	enthusiasm	proportion

1. She's a wonderfully creative dancer but she doesn't have the ________ of a truly great performer.
2. A higher ________ of men are willing to share household responsibilities than used to be the case.
3. Throughout history, people have been ________ by the question of whether there is intelligent life elsewhere in the universe.
4. One of the good things about teaching young children is their ________.
5. The findings show a ________ difference between the opinions of men and women.
6. I had to buy a new washing machine as it would not have been ________ to get it repaired.
7. The rock drawings ________ a variety of stylised human, bird and mythological figures and patterns.
8. Although the technology ________ in the UK, it has been developed in the US.
9. Watercolour painting began to ________ in Britain around 1750.
10. Her hopes of success in the race ________ last night as the weather became worse.

Banked cloze

Fill in the blanks by selecting suitable words from the word bank. You may not use any of the words more than once.

A. effective	F. garments	K. executes
B. pliable	G. absence	L. mimic
C. affected	H. contrast	M. existing
D. fosters	I. waning	N. innovative
E. embroidery	J. embroider	O. automatic

Historically the embroidery had, first of all, a decorative character. Nowadays this essential property favourably affected its use in advertising branch. We frequently see the 1. ________bearing advertising-informational character, on clothing objects and other ready-made 2. ________ . Embroidery can be an 3. ________ advertising-informational means, at the same time without loosing its artistic expressiveness. It is frequently used at trade marks plotting on 4. ________ material like the cloth, leather, felt, as well as on complete products, clothes, etc. At embroidery it is possible to reproduce trade marks of all colour combination. The trade mark, plotted on by this method, frequently turns into fashionable attribute of article and becomes an organic element of its composition. It is possible to 5. ________ on different types of clothes, including uniform, working clothes, etc. The embroidery as characteristic index to any current of activity is widely used in many countries. Very often the firm workers bear the differential signs of their company. It can be a logotype on the head gear, on the front side of the sweater or on the back side of the jacket.

Machine embroidery is a special type of embroidery, performed by special 6. ________ machines. The machine following the given program, automatically 7. ________ the embroidery on material. In this way, the main features of machine embroidery are the 8. ________ of manual labour at the stage of outright embroidery, and the possibility of embroidery samples duplication. The machine embroidery is a recent kind of activity, in 9. ________ to manual embroidery,

10. ________ for thousands of years.

Expressions in use

Fill in the blanks with the expressions given below. Change the form when necessary. Each expression can be used only once.

such as	on the scene	owe... to...	range from...to...
begin to	date back	conform to	part of

1. He's a page-turner, and his extraordinary brilliance as a critic is really just __________ the experience of reading him.
2. Before buying the baby's car seat, make sure that it __________ the official safety standards.
3. Over 75% of Mexico's forests, which __________ temperate spruce and fir __________ tropical rainforest, are controlled by ejidos or indigenous groups.
4. The President believes that we've got to do that and __________ add jobs based on a new foundation that doesn't depend on the bubble-and-bust economy that we've relied on for a long, long time.
5. That money is to cover costs __________ travel and accommodation.
6. I called the police and they were __________ within minutes.
7. This tradition __________ to medieval times.
8. The city essentially __________ its fame and beauty __________ the Moors who transformed it into the Muslim capital of Spain.

Translation

I. Translate the following paragraph into Chinese.

Su embroidery has a history of more than 2,600 years. In the Song Dynasty, Su embroidery had reached a large scale. There were embroidered clothes workshop, stores and alleys all over the city. In the Ming Dynasty, Su embroidery had developed its own characteristic and had a wide influence on other kinds of embroideries. The Qing Dynasty was a prosperous period of Su embroidery. At that time, most of the silk embroidery works used in royal families were produced in Suzhou. There were many skillful craftsmen in folk society. Suzhou artists are able to use more than 40 ways of needling and 1000 different types of threads to make flowers, birds, animals and even gardens on a piece of cloth. Su embroidery features a strong, folk flavour and its weaving techniques are characterised by the following: the product surface must be flat; the rim must be neat; the needle must be thin; the lines must be dense; the colour must be harmonious and bright and the picture must be even. Su embroidery products fall into three major categories: costumes, decorations for halls and crafts for daily use, which integrate decorative and practical values. Double-sided embroidery is an excellent representative of Su embroidery. The best-known work is an embroidered cat with bright eyes and fluffy hair, looking vivid and lifelike.

II. Translate the following paragraph into English.

中国的纺织技术历史悠久。早在原始社会时期，古人为了适应气候的变化，已懂得就地取材，利用自然资源作为纺织的原料，并懂得制造简单的纺织工具。中国机具纺织起源于新石器时期的纺轮。西周时期出现具有传统性能的纺车，汉代广泛使用提花机，唐代以后中国纺织机械日趋完善，大大促进了纺织业的发展。古今纺织工艺的发展取决于纺织原料，因此，原料在纺织技术中具有重要的地位。

Paragraph Writing

How to develop a paragraph—Description

Descriptive writing is a literary device in which the author uses details to paint a picture with their words. This process will provide readers with descriptions of people, places, objects, and events through the use of suitable details. The author will also use descriptive writing to create sensory details as a means of enhancing the reading experience. If done effectively, the reader will be able to draw a connection through the use of sensory details that include seeing, hearing, smelling, touching, and tasting. Through careful choice of words and phrasing, descriptive writing is vivid and detailed.

A good descriptive paragraph should be **concrete**, **evocative** and **plausible**. It should offer specifics the reader can envision, unite the concrete image with phrasing that evokes the impression the writer wants the reader to have, and constrain the concrete, evocative image to suit the reader's knowledge and attention span.

Here are some techniques that you can utilise and apply to your own descriptive writing.

1. **Begin with your topic sentence**: This sentence holds the key idea of the whole paragraph or states what the paragraph is all about.

2. **Use of words**: In most cases, you'll use adjectives to make your writing more detailed for the reader. This process will allow the reader to create a mental image through the use of your word choice. However, any unnecessary adjectives, adverbs, or clichéd figures of speech should be deleted. For example, try to compare the following two sentences:

a. The dog sniffs around.

b. The big brown dog sniffed around the red rose bushes in the front yard.

The use of "big," "brown," "red rose bushes," and "front yard" assists the reader in visualising the event and what the dog looks like.

3. **Use of sensory information**: By using the five senses of taste, smell, hearing, vision, and touch, you are creating an opportunity for the reader to develop an emotional connection to your writing.

4. **Use of figurative language**: There are several figurative languages that you can utilise in a descriptive paragraph such as metaphor, simile, personification, exaggeration, etc. Moreover, make sure to apply them correctly and do not overuse these tools.

Read the following examples carefully and pay attention to the adjectives, adverbs, and verbs appeared to experience and consider how the writers describe the weather and the characters to readers.

Example 1: The following paragraph is an excerpt from *Jamaica Inn* by Daphne du Maurier:

It was a cold grey day in late November. The weather had changed overnight, when a backing wind brought a granite sky and a mizzling rain with it, and although it was now only a little after two o'clock in the afternoon the pallor of a winter evening seemed to have closed upon the hills, cloaking them in mist.

Example 2: Here is Virginia Woolf, describing Mrs. Ramsay's husband's friend, Charles Tansley, in *To The Lighthouse*:

(Mrs. Ramsay) looked at him. He was such a miserable specimen, the children said, all humps and hollows. He couldn't play cricket; he poked; he shuffled. He was a sarcastic brute, Andrew said. They knew what he liked best—to be for ever walking up and down, up and down, with Mr. Ramsay, saying who had won this, who had won that ...

Example 3: The paragraph below shows how the Victorian author Charles Dickens depicts the boastful, self-important Mr. Bounderby in *Hard Times*:

He was a rich man: banker, merchant, manufacturer, and what not. A big, loud man, with a stare, and a metallic laugh. A man made out of coarse material, which seemed to have been stretched to make so much of him ... A man who was always proclaiming, through that brassy speaking-trumpet of a voice of his, his old ignorance and his old poverty. A man who was the Bully of humility.

Choose one of the following topics and write a descriptive paragraph of 80–100 words.

1. Describe a person you respect most.
2. Describe one of the impressive places you have visited.
3. Describe the appearance of a traditional handicraft.

Text B

The Craft of Lacemaking: Origins and Development

1 When did lace originate? Although no **definite** date can be given for the "invention" of lace, it is most likely that what we now regard as lace **arose** in the early sixteenth century. Open woven fabrics and fine nets that had a lace-like effect are known to have **existed** for centuries, but their techniques did not contribute to those developed for the great European laces. Early **references** to "lace" in English texts almost certainly refer to "ties", as this was the **primary** meaning of the word lace until well into the seventeenth century.

2 There is **pictorial** evidence from the late fifteenth century of simple **plaited** laces used on **costume**, and this is **consistent** with the statement by the author of a **bobbin lace** pattern book—*Nüw Modelbuch*—printed in Zurich in 1561, that lace was brought to Zurich from Italy in about 1536. What is certainly true is that the second half of the sixteenth century saw the rapid development of lace as an **openwork** fabric, created with a **needle** and single **thread** (needle lace) or with **multiple** threads (bobbin lace).

3 Bobbin lace **evolved** from braids and **trimmings** worked in colourful silks and silver-gilt threads and used as surface decoration for both dress and furnishings. Three forms of embroidery provided the origins of needlelace: (1) little **loops** and **picots** decorating the **collar** and **cuff edges** of shirts and **smocks**; (2) open-work seaming, linking widths of fabric; and (3) cutwork. Cutwork started as decorative stitching worked within small spaces cut out of linen. As the spaces became larger, leaving only a **grid** of the original threads, elaborate **geometric** patterns could be worked (known

as Reticella). In time, instead of cutting out expensive fabric, foundation threads were **couched** on to a **temporary** backing—usually **parchment**—and true needlelace lace was born. Designs were then able to break away from the geometric forms imposed by working within fabric, and the lace known as Punto in Aria (stitches in the air) was born.

4 Bobbin lace is generally quicker to work than needlelace, and skilled workers were soon able to copy needlelace designs. Details of such lace can be seen on hundreds of **portraits** from the 16th and 17th centuries.

5 Since lace evolved from other techniques, it is impossible to say that it originated in any one place, although the city whose name **was** first associated with lace is Venice. Venice was an important trading centre, and it was there that the first known lace pattern books were printed and in the early years the city certainly acted as a **hub** for the spread of lace knowledge. By 1600 high quality lace was being made in many centres across Europe including Flanders, Spain, France and England—women who were practised at other textile **crafts** seem to have picked up the new skills with relative ease.

6 Travelling **noblemen** and **intermarriage** between royal families ensured that new fashion ideas were **disseminated** widely: lace was traded and **smuggled** across borders. Lacemakers **displaced** by political upheavals often arrived as **refugees** in areas where there was already a lacemaking tradition and were able to enhance this with their own skills. And **enterprising** manufacturers of fashion for the **affluent** were constantly seeking **innovations** to secure and extend their position in the market.

7 Fashion has always driven lace production. Towards the end of the 16th century ruffs and standing collars demanded bold geometric needlelace. Through the early years of the 1600s these were gradually replaced by softer collars requiring many yards of relatively narrow linen bobbin lace. At the same time there was increasing demand for gold and silver lace to edge gloves, shoe roses, jackets and **sashes**, and also to provide surface

decoration for other garments. By the middle of the 17th century linen lace was again worn flat, and both needle and bobbin lace makers had refined their skills to produce some extremely intricate work, with the raised needlelace known as Gros Point and the flowing forms of Milanese bobbin lace being among the greatest achievements of the period.

8 Through the 18th century lace became increasingly delicate, often worked in extremely fine linen thread with increasing use of mesh grounds. French needle laces—Argentan and Alençon—and Flemish bobbin laces—Binche, Valenciennes, Mechlin—began to dominate the market, with items such as cravat ends and **lappets** used to **display** the wealth and **demonstrate** the good taste of the wearer.

9 The Industrial Revolution in Britain brought with it a **profound** change in lacemaking. The first machine lace was made towards the end of the 18th century, but it was not until 1809 that John Heathcoat was able to produce a wide net fabric that did not **unravel** when cut. This net became the basis for new laces such as Carrickmacross and Tambour (now classified as decorated nets), fabrics which were ideal for the light-weight dresses of the day. **Entrepreneurs** made constant improvements to the machines, first producing patterned nets, then increasingly complex designs, until by 1870 **virtually** every type of hand-made lace had its machine-made copy. Although there was a short period in the 1860s when bold laces such as Bedfordshire, Cluny and Yak (wool) were fashionable and could not yet be copied by machine, it became increasingly difficult for lacemakers such as those in Devon and the East Midlands to make a living from their work. In England most of the handmade lace industry had disappeared by 1900, although there were a number of small organisations (such as the North Bucks Lace Association) that supported lacemakers with patterns, training and an **outlet** for their work. There were a few parts of the world where hand-made lace was still produced for sale, but increasingly through the 20th century lacemaking became a craft undertaken for pleasure.

Notes

The Industrial Revolution The Industrial Revolution was a period of major industrialisation and innovation that took place during the late 1700s and early 1800s. The Industrial Revolution began in Great Britain and quickly spread throughout the world.

Zurich 苏黎世（瑞士北部城市）

Venice 威尼斯（意大利港口城市）

Flanders 佛兰德斯（历史地名，包括现比利时、荷兰以及法国的部分地区）

John Heathcoat 约翰·希斯科特（英国发明家）

New words and phrases

definite /ˈdefɪnət/ *a.* ① sure or certain; unlikely to change 肯定的；确定的；不会改变的 ② easily or clearly seen or understood; obvious 清楚的；明显的

arise /əˈraɪz/ *v.* (*formal*) to begin to exist or develop 出现；产生；发展

exist /ɪɡˈzɪst/ *v.* ① to be real; to be present in a place or situation 存在；实际上有 ② ~ (**on sth**) to live, especially in a difficult situation or with very little money（尤指在困境或贫困中）生活，生存

reference /ˈrefrəns/ *n.* [C, U] ~ (**to sb/sth**) a thing you say or write that mentions sb/sth else; the act of mentioning sb/sth 说到（或写到）的事；提到；谈及；涉及

primary /ˈpraɪməri/ *a.* ① main; most important; basic 主要的；最重要的；基本的 ② developing or happening first; earliest 最初的；最早的

pictorial /pɪkˈtɔːriəl/ *a.* ① using or containing pictures 用图片的；有插图的 ② connected with pictures 画片的；图画的

plait /plæt/ *v.* to twist three or more long pieces of hair, rope, etc. together to make one long piece 将（头发、绳子等）编成辫

costume /ˈkɒstjuːm/ *n.* [C, U] ① the clothes worn by people from a particular place or during a particular historical period（某地或某历史时期的）服装，装束 ② the clothes worn by actors in a play or film/movie, or worn by sb to make them look like sth else（戏剧或电影的）戏装，服装

consistent /kənˈsɪstənt/ *a.* ~ **with sth** in agreement with sth; not contradicting sth 与……一致的；相符的；符合的；不矛盾的

bobbin /ˈbɒbɪn/ *n.* [C] a small device on which you wind thread, used, for example, in a sewing machine 线轴；绕线筒

bobbin lace *n.* lace made with bobbins rather than with needle and thread (needlepoint lace); pillow lace 梭结花边

openwork /ˈəʊpənˌwɜːk/ *n.* ornamental work, as of metal or embroidery, having a pattern of openings or holes 有透孔的金属或刺绣织物

needle /ˈniːdl/ *n.* [C] ① a small thin piece of steel that you use for sewing, with a point at one end and a hole for the thread at the other 针；缝衣针 ② a long thin piece of plastic or metal with a point at one end that you use for knitting 编织针

thread /θred/	*n*. [U, C] a thin string of cotton, wool, silk, etc. used for sewing or making cloth（棉、毛、丝等的）线
multiple /ˈmʌltɪpl/	*a*. many in number; involving many different people or things 数量多的；多种多样的
evolve /iˈvɒlv/	*v*. ~ **(sth) (from sth) (into sth)** to develop gradually, especially from a simple to a more complicated form; to develop sth in this way（使）逐渐形成，逐步发展，逐渐演变
trimming /ˈtrɪmɪŋ/	*n*. [U] [C, usually pl.] material that is used to decorate sth, for example along its edges 装饰材料；镶边饰物
loop /luːp/	*n*. [C] ① a shape like a curve or circle made by a line curving right round and crossing itself 环形；环状物；圆圈 ② a piece of rope, wire, etc. in the shape of a curve or circle（绳、电线等的）环，圈
picot /ˈpiːkəʊ/	*n*. any of a pattern of small loops, as on lace 花边上饰边的小环
collar /ˈkɒlə(r)/	*n*. [C] the part around the neck of a shirt, jacket or coat that usually folds down 衣领；领子
cuff /kʌf/	*n*. [C] the end of a coat or shirt sleeve at the wrist 袖口
edge /edʒ/	*n*. [C] the outside limit of an object, a surface or an area; the part furthest from the centre 边；边缘；边线；边沿

smock /smɒk/ *n.* [C] ① a loose comfortable piece of clothing like a long shirt, worn especially by women（多为女性穿的）宽松式衬衫 ② a long loose piece of clothing worn over other clothes to protect them from dirt, etc. 罩衣；工作服

grid /ɡrɪd/ *n.* [C] a pattern of straight lines, usually crossing each other to form squares 网格；方格

geometric /ˌdʒiːəˈmetrɪk/ *a.* of geometry; of or like the lines, shapes, etc. used in geometry, especially because of having regular shapes or lines 几何（学）的；（似）几何图形的

couch /kaʊtʃ/ *v.* to embroider (a design) by laying down a thread and fastening it with small stitches at regular intervals 贴线缝绣

temporary /ˈtemprəri/ *a.* lasting or intended to last or be used only for a short time; not permanent 短暂的；暂时的；临时的

parchment /ˈpɑːtʃmənt/ *n.* ① [U] material made from the skin of a sheep or goat, used in the past for writing on 羊皮纸 ② [U] a thick yellowish type of paper 仿羊皮纸 ③ [C] a document written on a piece of parchment 羊皮纸文献

portrait /ˈpɔːtreɪt/ *n.* [C] ① a painting, drawing or photograph of a person, especially of the head and shoulders 肖像；半身画像；半身照 ② a detailed description of sb/sth 详细的描述；描绘

hub /hʌb/ *n.* [usually sing.] ~ **(of sth)** the central and most important part of a particular place or activity（某地或活动的）中心，核心

craft /krɑːft/ *n.* ① [C, U] an activity involving a special skill at making things with your hands 手艺；工艺 ② [sing.] all the skills needed for a particular activity 技巧；技能；技艺

nobleman /ˈnəʊblmən/ *n.* [C] a person from a family of high social rank; a member of the nobility 出身高贵的人；贵族成员

intermarriage /ˌɪntəˈmærɪdʒ/ *n.* [U] marriage between people from different social, racial, or religious groups 通婚

disseminate /dɪˈsemɪneɪt/ *v.* (*formal*) to spread information, knowledge, etc. so that it reaches many people 散布，传播（信息、知识等）

smuggle /ˈsmʌgl/ *v.* to take, send or bring goods or people secretly and illegally into or out of a country, etc. 走私；私运；偷运

displace /dɪsˈpleɪs/ *v.* ① to take the place of sb/sth 取代；替代；置换 ② to force people to move away from their home to another place 迫使（某人）离开家园 ③ to move sth from its usual position 移动；挪开；转移 ④ (*especially NAmE*) to remove sb from a job or position 撤职；免职；使失业

refugee /ˌrefjuˈdʒiː/ *n.* [C] a person who has been forced to leave their country or home, because there is a war or for political, religious or social reasons 避难者；逃亡者；难民

enterprising /ˈentəpraɪzɪŋ/ *a.* (*approving*) having or showing the ability to think of new projects or new ways of doing things and make them successful 有事业心的；有进取心的；有创业精神的

affluent /ˈæfluənt/	*a.* having a lot of money and a good standard of living 富裕的
innovation /ˌɪnəˈveɪʃn/	*n.* ① [U] the introduction of new things, ideas or ways of doing sth（新事物、思想或方法的）创造；创新；改革 ② [C] a new idea, way of doing sth, etc. that has been introduced or discovered 新思想；新方法
sash /sæʃ/	*n.* [C] a long strip of cloth worn around the waist or over one shoulder, especially as part of a uniform（尤指制服的）腰带，肩带，饰带
lappet /ˈlæpɪt/	*n.* a small hanging flap or piece of lace, etc., such as one dangling from a headdress 垂饰
display /dɪˈspleɪ/	*v.* ① to put sth in a place where people can see it easily; to show sth to people 陈列；展出；展示 ② to show signs of sth, especially a quality or feeling 显示，显露，表现（特性或情感等）
demonstrate /ˈdemənstreɪt/	*v.* ① ~ **sth (to sb)** to show sth clearly by giving proof or evidence 证明；证实；论证；说明 ② to show by your actions that you have a particular quality, feeling or opinion 表达；表露；表现；显露
profound /prəˈfaʊnd/	*a.* very great; felt or experienced very strongly 巨大的;深切的；深远的
unravel /ʌnˈrævl/	*v.* If you unravel threads that are twisted, woven or knitted, or if they unravel, they become separated.（把缠或织在一起的线）解开，拆散，松开
entrepreneur /ˌɒntrəprəˈnɜː(r)/	*n.* [C] a person who makes money by starting or running businesses, especially when this involves taking financial risks 创业者，企业家（尤指涉及财务风险的）

virtually /ˈvɜːtʃuəli/	*ad.* almost or very nearly, so that any slight difference is not important 几乎；差不多；事实上；实际上
outlet /ˈaʊtlet/	*n.* [C] ① a way of expressing or making good use of strong feelings, ideas or energy（感情、思想、精力发泄的）出路；表现机会 ② a shop/store or an organisation that sells goods made by a particular company or of a particular type 专营店；经销店
regard sb/sth as sth	to think about sb/sth in a particular way 把……视为……；看待
contribute to sth	to be a factor or catalyst for a particular occurrence or event 是……的原因之一
refer to sb/sth	to describe or be connected to sb/sth 描述；涉及；与……相关
be consistent with	in agreement with sth; not contradicting sth 与……一致的；相符的；符合的；不矛盾的
evolve from	to develop gradually, especially from a simple to a more complicated form; to develop sth in this way（使）逐渐形成，逐步发展，逐渐演变
break away from sb/sth	move away or escape suddenly; withdraw from an organisation or a larger group, etc. 突然挣脱；逃脱；脱离
be associated with sth	If problems or dangers are associated with a particular thing or action, they are caused by it. 与……有关
pick up	get to know or become aware of, usually by chance rather than by making a deliberate effort（偶然）得到，听到，学会

replace sb/sth with/by sb/sth	to remove sb/sth and put another person or thing in their place 用……替换；以……接替
at the same time	sumultaneously; at once 同时
a number of	a quantity of 若干，许多

Reading Comprehension

Understanding the text

Choose the best answer to each of the following questions.

1. When did lace originate?
   A. In the early 16th century.
   B. In the late 16th century.
   C. In the early 17th century.
   D. In the late 17th century.
2. What is bobbin lace used as?
   A. It is used as surface decoration for furnishings.
   B. It is used as surface decoration for both dress and furnishings.
   C. It is used as surface decoration for both gowns and pants.
   D. It is used as surface decoration for both gowns and furnishings.
3. Why is the city of Venice associated with lace?
   A. Because lace originated in Venice and Venice was the only trading centre of lace.
   B. Because lace was first traded in Venice and Venice acted as a hub for the spread of lace knowledge.
   C. Because Venice was an important trading centre of lace and acted as a hub for the spread of lace knowledge.
   D. Because lace originated in Venice and Venice used to be the only trading centre of lace.
4. Which one of the following is not used to display the wealth and demonstrate the good taste of the wearer in the 18th century?
   A. Gros Point.

B. Argentan.

C. Alençon.

D. Valenciennes.

5. How did the Industrial Revolution in Britain bring a profound change in lacemaking?

A. The first machine lace was made.

B. New laces such as Carrickmacross and Tambour appeared.

C. It became increasingly difficult for lacemakers to make a living from their work.

D. All of the above.

Mind mapping

Read Text B carefully and try to find out the key information to complete the evolution of lace according to the timeline.

the late 15th century: 1. ________

↓

the second half of the 16th century: 2. ________

↓

1600: 3. ________

↓

the end of the 16th century: 4. ________

↓

the early years of the 1600s: 5. ________

↓

the middle of the 17th century: 6. ________

↓

the 18th century: 7. ________

↓

the end of the 18th century: 8. ________

↓

the 1860s: 9. ________

↓

1870: 10. ________

↓

1900: 11. ________

↓

the 20th century: 12. ________

Language Enhancement

Words in use

Fill in the blanks with the words given below. Change the form when necessary. Each word can be used only once.

disseminate	definite	demonstrate	pictorial	evolve
temporary	innovation	profound	consistent	arise

1. We'll have to wait until the new year before we can make any ________ plans.
2. His reputation for carelessness was established long before the latest problems ________ .
3. The exhibition is a ________ record of the town in the 19th century.
4. What the witness said in court was not ________ with the statement he made to the police.
5. Most languages are constantly ________ and changing, which is what keeps them alive.
6. The salaries of ________ employees ought to be brought into line with those of permanent staff.
7. One of the organisation's aims is to ________ information about the disease.
8. To go in for technical ________, one must have the fearless spirit of a pathbreaker.
9. It is impossible to ________ conclusively that the factory is responsible for the pollution.
10. They have found that music lessons can produce ________ and lasting changes that enhance the general ability to learn.

Expressions in use

Fill in the blanks with the expressions given below. Change the form when necessary. Each expression can be used only once.

pick up	be consistent with	a number of	evolve from
refer to	contribute to	be associated with	at the same time

1. Medical negligence was said to have ____________ her death.
2. The term "Arts" usually ____________ humanities and social sciences.
3. The results are entirely ____________ our earlier research.
4. It is interesting that Chinese characters ____________ pictures and signs.
5. But ____________ we might have very different experiences and that was amplified with students coming from different cultures altogether.
6. There were __________ major causes of the financial crisis and panic of 2008.
7. The cancer risks ____________ smoking have been well documented.
8. She ____________ Spanish when she was living in Mexico.

Sentence structure

I. Complete the following sentences by translating the Chinese into English, using "it was not until... that..." structure.

Model: ________________that did not unravel when cut. [直到1809年，约翰·希思科特（John Heathcoat）才能够生产出一种宽幅的网状织物。]

→ It was not until 1809 that John Heathcoat was able to produce a wide net fabric that did not unravel when cut.

1. It allowed women to take the university exams in 1881 in America, but ____.

（直到1948年，女性才可以获得学位。）

2. Architectural acoustics is acknowledged as an important branch of acoustical engineering. However, ____.

（直到二十世纪初，建筑声学才正式成为一个科学领域。）

3. Bach died in 1750, while ____.

（直到19世纪早期，他的音乐天赋才充分得到赏识。）

II. Rewrite the following sentences by using the independent structure "with + n. + v.-ed".

Model: French needle laces began to dominate the market, and items such as cravat ends and lappets are used to display the wealth and demonstrate the good taste of the wearer.

→French needle laces began to dominate the market, with items such as cravat ends and lappets used to display the wealth and demonstrate the good taste of the wearer.

1. English class is getting much more meaningful and interesting since textbooks are designed and used in conjunction with classroom teaching.

   ____

   ____.

2. The patient with thalassemia (地中海贫血) is getting much better through the successful application of gene therapy.

   ____

   ____.

3. The relationship between them is restored after they clear up misunderstandings.

   ____

   ____.

Chinese Batik: Wax Printing

1 Wax printing, also known as batik, is one of the three ancient Chinese handicraft methods of producing dyed, multi-coloured textiles via a process which prevents the dye from reaching certain (chosen) parts of the fabric (the other two dyeing methods are bandhnu and calico). It is believed that wax printing existed in China as early as during the late Qin

or early Han Dynasty, though widespread knowledge the existence of Chinese batik as a finished product first occurred during the Tang Dynasty, when batik became yet another "Silk Road" commodity that was exported to Europe and elsewhere.

2 Wax printing, alternatively batik printing, is a mechanical dye-blocking method whereby hot melted wax is applied, often in the form of a geometric pattern or an artistic representation (anything from a flower to a human face), to a chosen part of the fabric, then when the wax has dried sufficiently, the fabric is dyed in a cold-water vat of soluble dye. When the dyeing process is finished and the fabric has been allowed to dry completely, the fabric is then washed in hot water, which dissolves the wax, and the finished product is a piece of fabric with patterns, designs, images, etc., in a contrasting colour to the dyed, or background, colour.

Categories

3 In general, though those who expound on thc subject tend to present the case in a much more nuanced light than is justified, the methods of producing multi-coloured textiles via one or more dyeing processes can be divided into two overarching categories: mechanical and chemical. All three of the ancient Chinese handicraft methods of producing dyed, multi-coloured textiles fall into the former category.

4 The mechanical method of partial dyeing is essentially aimed at physically blocking, in some cases, only impeding or reducing, the dye's access to the targeted parts of the fabric, whereas the chemical (aka reactive) method is to use a chemical agent to temporarily render an insoluble dyeing agent soluble, so that it may be absorbed into the fibers of the targeted parts of the fabric, where, once dried (i.e., oxidised), the absorbed dye in this part of the fabric will return to its insoluble state, so that when one later dyes the entire piece of fabric in a vat containing a different coloured dye, the insoluble dye prevents the second (soluble) dye from being absorbed into

the fibers in those parts of the fabric where the insoluble dye has been absorbed.

5 Naturally, one can create multiple shades of these colour patterns by dyeing more than once—using the same or a different (darker) colour of dye—where more hot wax in the targeted area (not necessarily the same areas as in the first dyeing process) will preserve the existing colour, while the rest of the fabric is impacted by the second (third, fourth, etc.) dyeing process.

Batik Skills

6 Applying colour via the wax printing method is, if not an art, at least a craft which requires a high degree of skill, not least because the hot wax must be applied in small amounts, quickly—otherwise the wax will cool and not be properly absorbed into the fibers of the fabric—and often deftly, otherwise the image's contours will lack sharpness. The nature of the design or image will dictate the type of "spatula" used: either broad or finely tapered.

7 The batik artist, in applying the hot wax that will prevent the dye from penetrating the fibers of the fabric in the targeted areas, thus produces a negative image (think of the image of a face), in the sense that the areas where the wax is applied represent reflective surfaces (think of the cheeks) while the areas where the wax is not applied represents shadowy surfaces (think of the recessed areas around the eyes). This requires not only a good grasp of how to draw an object (almost all batik was traditionally done in freehand, not with the help of design aids such as stencils or pre-drawn contours), but also of how to draw it "in reverse".

Batik Communities in China

8 In China, it is believed that the art of wax printing was widespread, but this being a tradition that is passed on from generation to generation, it disappeared in many, if not most, communities in China as Chinese society

evolved and batik was abandoned, for whatever reasons. Today there are only two communities, the Zhuang and the Miao ethnic groups, living in small enclaves in Guizhou, Guangxi, Sichuan and Yunnan Provinces, which have preserved the ancient tradition of batik, or wax printing. This comes as no surprise, as many oral traditions as well are preserved only in smaller ethnic communities. That China highlights ethnic minority tours to Guizhou offers tourists a great chance to explore the ancient culture of the Miao and Dong ethnic people in Guizhou, including their centuries-old batik culture.

9 The Zhuang ethnic minority have a preference for dyed-blue fabric with white flower blossoms that seem to leap out of the cloth. The cloth is first bleached so as to render it pasty-white, allowed to dry thoroughly, then the hot wax is applied in the shape of a flower blossom. The rest is as described above. The basic batik method used by the Zhuang ethnic minority is also followed by the Miao, who similarly work with bleached white cloth, except that the Miao employ a wider variety of representational and non-representational images in their batik.

10 The Zhuang and the Miao not only make batik for their own use, but produce batik for sale. Tourists who visit these ethnic areas will find wax-print articles ranging from home decor items (curtains, cushions, tablecloths, wall hangings, etc.) to personal items such as handbags, dolls, and clothing.

Unit 4

Prominent Figures

Design is in everything we make, but it's also between those things. It's a mix of craft, science, storytelling, propaganda, and philosophy.

—Erik Adigard

A designer knows he has achieved perfection not when there is nothing left to add, but when there is nothing left to take away.

—Antoine de Saint-Exupéry

Pre-Reading Activities

1. Listen to an interview with Vera Wang. Fill in the blanks based on what you hear.

(1) Vera Wang possesses the ability to realize that she can not only be a ________; she has to ________ and find her own ________ .

(2) She once was a ________ skater and ________ 13 years in her life skating 8 hours a day.

(3) In her opinion, work seems two sides. The better one gets at, the more he becomes, the more ________ and more ________ he'll have about himself.

(4) She believes ________. If one is not able to get where he wishes to go, but he has to also have some ________ and a sense of ________.

2. Listen to the interview again and answer the following questions.

(1) According to Vera Wang, what are part of heritage in the Chinese culture?

(2) Are there any other designers familiar to you? Share their stories to your group members.

Mid-Century Textile Designer Lucienne Day

1 Of all the talented women textile designers of post-war Britain, Lucienne Day's influence is the most **far-reaching**.

2 Graduating from the Printed Textiles department of the Royal College of Art in 1940, the effects of the Second World War **initially** limited the prospects of design work, and Lucienne Day **supplemented** her design career with teaching roles. However, as the **restrictions** of the war began to lift, she quickly built on her relationships with existing manufacturing clients to produce modern **furnishing** textile designs. Along with her husband, furniture-designer Robin Day, she promoted modern living and embodied the image of the newly styled professional designer.

3 This **deferred launch** of Lucienne Day's career **coincided** with a major governmental **initiative** to **boost** the nation's industrial production by **elevating** the status, training and **consequently** the output of British designers. Both Lucienne and Robin fulfilled the brief perfectly: **ambitious**, highly talented and with a **committed** vision of the life-changing potential design could bring.

4 Although Lucienne was already creating **progressive** designs for the dress industry, her first commercially produced furnishing textiles were not **overly avant-garde**, loosely acknowledging the long-favoured tradition of floral **chintz** for home furnishing fabrics. Day believed that a designer must be practical and meet market needs—her early designs were well-judged to appeal to traditional consumer tastes. Their success brought her further **commissions** and set her on the path to become a sought-after **freelance** designer.

"Calyx" and the Festival of Britain

5 In 1951, her husband **secured** the commission to design all the seating for the Royal Festival Hall, the newly-built auditorium and concert as part of the Festival of Britain—a major exhibition intended to **showcase** Britain and its achievements. He also designed room settings for the Festival's Homes and Gardens **Pavilion**, which showcased his own **contemporary** furniture.

6 For these room settings he needed a wallpaper and furnishing fabric that

would sit comfortably with his modern furniture. Provence wallpaper, a previous design by Lucienne Day, was chosen for one room to **illustrate** an affordable **interior**. Lucienne also **conceived** a new textile design, "Calyx", intending to **complement** Robin's more costly living/dining room. Heals were initially hesitant to support the design due to its **radical** nature—but they eventually agreed.

7 "Calyx" uses a very traditional source of **inspiration**, **botanical** form, but the plant motifs are here stylised almost to the point of **abstraction**, and are linked with **diagonal** and **vertical** thin solid and dotted lines, suggesting flower **stalks**. The design embodies the springing energy of new growth, perfectly **encapsulating** the spirit of the Festival and general social **optimism** of the time.

8 Lucienne Day's new design turned out to have great market appeal and became one of many commercial successes in a long-standing partnership between Heals and the designer. It is now recognised as a **seminal** piece of British post-war design.

Inspiration and Influences

9 During her training at the Royal College of Art, Lucienne spent many hours in the galleries of the Victoria and Albert Museum. Here she found inspiration for her degree show, in the form of a Chinese sculpture of a horse. In her design, *Horse's Head*, the horse is refined to a simple motif, which is **alternated** with **fluidly** drawn squares, hand-printed onto linen. The Museum's collection provided inspiration again for *Script* (1956). Botanical form was an important source of inspiration throughout Lucienne Day's career—her treatment of this theme provides much of the visual delight and innovation in her designs.

10 In *Trio* (1954), small groups of flower-like forms stand upright against a striped background. In *Herb Antony* (1956), plant elements are **transformed** into graceful line-drawn forms that take on a **unique** character, **hovering** between reality and imagination. In her highly thoughtful approach to

design, Lucienne also benefitted from an **appreciation** of contemporary fine art as well as possessing something of an artist's **sensibility** herself. The absorbed influence of the work of artists such as Joan Miró and Paul Klee can sometimes be **perceived**—and in many cases Lucienne Day's work can be seen to **anticipate** or equal fine art influences. *Graphica* (1953) for example, is a highly accomplished minimal geometric abstraction and *Causeway* (1967), a brilliant execution of colour block contrast.

11 Regardless of the increasing celebrity status, Lucienne remained committed to answering the material needs and demands of the consumer market. While the large scale, printed linen "Calyx" worked well in spacious interiors, hung from ceiling to floor, Heals also requested designs to suit smaller homes, and with a more modest price tag. Enthusiastic to produce affordable textiles, she used newly developed and cheaper man-made fibres, such as rayon. She designed "Miscellany", "Quadrille" and "Palisade" for British Celanese—a company that specialised in producing man-made fabrics from cellulose fibres, which were affordable as well as being soft and durable. In the later 1960s, her designs moved with the spirit of the times to larger geometric based pattern in **bold** colours, illustrating her ability to adapt and evolve.

12 Though Lucienne Day is best-known for her patterns for furnishing fabrics, she also produced designs for many other applications. Dress fabrics were an important part of her design practice in the early post-war period. With some companies, she **collaborated** to produce several successful collections of wallpapers. Diabolo was one of three wallpapers launched at the Festival of Britain in 1951. During the 1950s and 1960s she produced a substantial body of carpet designs, including Tesserae which won a Design Centre Award in 1957.

13 In the late 1970s, with a **steadfast** belief in modernism, Lucienne Day started creating one-off compositions in silk. Using a construction technique derived from traditional **patchwork**, her "Silk Mosaics" are composed of 1 cm squares and strips of coloured silk. The design emerges through carefully **juxtaposed** blocks of colours and weave textures. These creations occupied the decades of the 1980s and 1990s.

14 In brief, with her independent spirit, Lucienne Day was part of a new generation of designers who came to the forefront of the British textile industry in the immediate post-war period. By setting a new standard in originality, quality and her breakthrough pattern "Calyx", she maintained her position as one of the most versatile and influential designers both in Britain and abroad for many years.

Notes

The Royal College of Art Established in 1837 and located in the heart of London, the Royal College of Art is the world's oldest art and design university in continuous operation, with a tradition of innovation and excellence in creative education.

Calyx A stylised floral design mixing muted and acid colours, it represented a radical new aesthetic in pattern design. As one of Lucienne Day's most famous textiles, it was originally shown at the Festival of Britain in 1951, and is considered her breakthrough design.

Robin Day 罗宾·戴（20 世纪最具影响力的英国家具设计师之一，最知名的设计作品是 1962 年的聚丙烯塑胶椅）

Provence 普罗旺斯

Victoria and Albert Museum 维多利亚和阿尔伯特博物馆（世界上最大的装饰艺术和设计博物馆）

Joan Miro 胡安·米罗（西班牙著名画家、雕塑家、陶艺家、版画家，超现实主义风格的代表人物。）

Paul Klee 保罗·克利（瑞士著名画家，其作品以对色彩的感悟为特征，在抽象与装饰性图形之间自由转换。）

Miscellany 大杂烩；杂集

Quadrille 规则小方格

Palisade 木栅栏

British Celanese 英国塞拉尼斯公司

New words and phrases

far-reaching /ˌfɑː ˈriːtʃɪŋ/ *a.* having a great influence or effect on many people or things 影响深远的；波及广泛的

initially /ɪˈnɪʃəli/ *ad.* at the beginning 最初；一开始

supplement /ˈsʌplɪmənt/ *v.* to add something to something to make it larger or better 增补；增加；补充

n. a thing that is added to something else to improve or complete it 增补（物）；补充（物）

restriction /rɪˈstrɪkʃn/ *n.* ① [C] ~ **(on sth)** a rule or law that limits what you can do or what can happen 限制规定；限制法规 ② [U] the act of limiting or controlling sb/sth 限制；约束

furnishing /'fɜːnɪʃɪŋ/ *n.* [C] ① accessory wearing apparel 服饰 ② the furniture, curtains, and other decorations in a room or building 家具；室内陈设

deferred /dɪˈfɜːd/ *a.* postponed 推迟的；延期的

launch /lɔːntʃ/ *n.* [usually sing.] making a start in a career or vocation 开始，创办

v. ① to make a product available to the public for the first time （首次）上市，发行，推出 ② send (a missile, satellite or spacecraft) on its course or into orbit 发射（人造卫星、导弹或航天器）

coincide /ˌkəʊɪnˈsaɪd/ *v.* ① (of two or more events) to happen at or near the same time（几乎）同时发生 ② (of ideas, opinions, etc.) to be the same or very similar 相同；相符；极为类似

initiative /ɪˈnɪʃətɪv/ *n.* ① [C] an important new plan or statement to achieve a particular aim or to solve a particular problem 倡议；新举措 ② [U] the ability to decide and act on your own without waiting for somebody to tell you what to do 主动性；积极性；自发性

boost /buːst/ *v.* to improve or increase something 改善；提高；增强；推动

elevate /ˈelɪveɪt/ *v.* ① to make someone or something more important or to improve something 提升；提高；改进 ② to raise something or lift something up 抬高；使上升；举起

consequently /ˈkɒnsɪkwəntli/ *ad.* as a result; therefore 因此；所以

ambitious /æmˈbɪʃəs/ *a.* having a strong wish to be successful, powerful, or rich 有抱负的；志向远大的；雄心勃勃的

committed /kəˈmɪtɪd/ *a.* willing to work hard and give your time and energy to sth; believing strongly in sth 尽心尽力的；坚信的；坚定的

progressive /prəˈgresɪv/ *a.* ① in favour of new ideas, modern methods and change 进步的；先进的；开明的 ② happening or developing steadily 稳步的；逐步的；稳定发展的

overly /ˈəʊvəli/ *ad.* too; very; excessively 很；十分；过于；过度地

avant-garde /ˌævɒŋˈɡɑːd/ — *a.* (ideas, styles, and methods) being very original or modern in comparison to the period in which they happen 前卫的；先锋派的

chintz /tʃɪnts/ — *n.* [U, C] a type of shiny cotton cloth with a printed design, especially of flowers, used for making curtains, covering furniture, etc. 轧光印花棉布（用于制作窗帘、家具套等）

commission /kəˈmɪʃn/ — *n.* ① [C] a piece of work that someone is asked to do and is paid for 委托任务 ② [C, U] an amount of money that is paid to sb for selling goods and which increases with the amount of goods that are sold 佣金；回扣

freelance /ˈfriːlɑːns/ — *a.* doing particular pieces of work for different organisations, rather than working all the time for a single organisation 从事自由职业（的）；作为自由职业者（的）

secure /sɪˈkjʊə(r)/ — *v.* ① **~ sth (for sb/sth) | ~ sb sth** (*formal*) to obtain sth, often after a lot of effort 获得；争取到 ② **~ sth (against sth)** to protect sth so that it is safe and difficult to attack or damage 保护；保卫；使安全

a. feeling happy and confident about yourself or a particular situation 安心的；有把握的

showcase /ˈʃəʊkeɪs/ — *v.* to show the best qualities or parts of sth 展示……的优点；充分展示

n. [C] a container with glass sides in which valuable or important objects are kept so that they can be looked at without being touched, damaged, or stolen （玻璃）陈列柜，展示橱

pavilion /pəˈvɪliən/

n. [C] ① a summer house or other decorative building used as a shelter in a park or large garden（公园或大花园中的）凉亭，亭子；阁 ② a temporary building, stand, or other structure in which items are displayed by a trader or exhibitor at a trade exhibition 临时展览馆，临时展棚，临时展摊

contemporary /kənˈtemprəri/

a. ① belonging to the present time 当代的；现代的 ② belonging to the same time 属同时期的；同一时代的

illustrate /ˈɪləstreɪt/

v. to show the meaning or truth of sth more clearly, especially by giving examples（尤指用例子）说明，阐明

interior /ɪnˈtɪəriə(r)/

n. [C] ① an artistic representation of the inside of a building or room（建筑物或房间的）内景 ② the inner or indoor part of sth, especially a building; the inside（尤指）建筑物内部，屋内，室内；里面

a. connected with the inside part of sth 内部的；里面的

conceive /kənˈsiːv/

v. ① to invent a plan or an idea 构想，想出，设想出（计划或主意）② to become pregnant 怀孕；受孕；怀（胎）

complement /ˈkɒmplɪm(ə)nt/

v. add to sth in a way to make it seem better or more attractive 补充；补足；使完善；为……增色，衬托

n. [C] ① ~ **(to sth)** a thing that adds new qualities to sth in a way that improves it or makes it more attractive 补充物；补足物；衬托物 ② a part of a clause that usually follows the verb in English and adds more information about the subject or object 补语

radical /ˈrædɪkl/ *a.* ① new, different and likely to have a great effect 全新的；不同凡响的 ② concerning the most basic and important parts of sth; thorough and complete 根本的；彻底的；完全的 ③ in favour of thorough and complete political or social change 激进的；极端的

inspiration /ˌɪnspəˈreɪʃn/ *n.* ① [U] the process that takes place when someone sees or hears sth that causes them to have exciting new ideas or makes them want to create sth, especially in art, music or literature 灵感 ② [C] a person or thing that is the reason why sb creates or does sth 启发灵感的人（或事物）；使人产生动机的人（或事物）

botanical /bəˈtænɪkl/ *a.* of or relating to botany （与）植物（有关）的；（与）植物学（有关）的

abstraction /æbˈstrækʃn/ *n.* ① [U] freedom from representational qualities in art （艺术上的）抽象 ② [U, C] a general idea not based on any particular real person, thing or situation; the quality of being abstract 抽象概念；抽象 ③ [U] the state of thinking deeply about sth and not paying attention to what is around you 出神；心神专注

diagonal /daɪˈægənl/ *a.* at an angle; joining two opposite sides of sth at an angle 斜线的；对角线的

vertical /ˈvɜːtɪkl/ *a.* standing or pointing straight up or at an angle of 90° to a horizontal surface or line 竖直的；垂直的；立式的

stalk /stɔːk/ — *n.* [C] the main stem of a plant, or the narrow stem that joins leaves, flowers, or fruit to the main stem of a plant（植物的）茎，杆；（叶或果实的）柄，蒂；（花）梗

encapsulate /ɪnˈkæpsjuleɪt/ — *v.* ① express the essential features of (sb or sth) succinctly 简要描述；概括；浓缩 ② enclose (sth) in or as if in a capsule 用胶囊（或囊状物）装；封装

optimism /ˈɒptɪmɪzəm/ — *n.* [U] ~ **(about/for sth)** a feeling that good things will happen and that sth will be successful; the tendency to have this feeling 乐观；乐观主义

seminal /ˈsemɪnl/ — *a.* containing important new ideas and having a great influence on later work 具有开拓性的；有深远影响的

alternate /ˈɔːltəneɪt/ — *v.* ~ **(with sth)** (of things or people) to follow one after the other in a repeated pattern 交替；轮流

fluidly /ˈfluːɪdli/ — *ad.* not in a settled or stable way 不固定地，不稳定地，易变地

transform /trænsˈfɔːm/ — *v.* ① to change the form of sth 使改变形态 ② to completely change the appearance or character of sth, especially so that it is better 使彻底改观；使大变样

unique /juˈniːk/ — *a.* ① very special or unusual 独特的；罕见的 ② being the only one of its kind 唯一的；独一无二的

hover /ˈhɒvə(r)/ — *v.* ① to remain in one place in the air 盘旋，翱翔 ② to stand somewhere, especially near another person, eagerly or nervously waiting for their attention 徘徊，守候

appreciation /əˌpriːʃiˈeɪʃn/ *n.* [U] ① recognition and enjoyment of the good qualities of sb or sth 赏识；欣赏 ② ~ **of sth** a full or sympathetic understanding of sth, such as a situation or a problem, and of what it involves 理解；体谅；同情 ③ ~ **(of/for sth)** the feeling of being grateful for sth 感激；感谢

sensibility /ˌsensəˈbɪləti/ *n.* [U, C] the ability to experience and understand deep feelings, especially in art and literature（尤指文艺方面的）感受能力，鉴赏力，敏锐

perceive /pəˈsiːv/ *v.* to notice or become aware of sth 注意到；意识到；察觉到

anticipate /ænˈtɪsɪpeɪt/ *v.* to imagine or expect that something will happen 预期，期望；预料

bold /bəʊld/ *a.* ① strong in colour or shape and very noticeable 醒目的；显著的；色彩艳丽的；轮廓清晰的 ② not frightened of danger 勇敢的，无畏的 ③ in a thick, dark type 粗体的；黑体的

collaborate /kəˈlæbəreɪt/ *v.* ① ~ **(with sb) (on sth)** | ~ **(with sb) (in sth/ in doing sth)** to work together with sb in order to produce or achieve sth 合作；协作 ② to help the enemy who has taken control of your country during a war 通敌；勾结敌人

steadfast /ˈstedfɑːst/ *a.* not changing in your attitudes or aims 坚定的；不动摇的

patchwork /ˈpætʃwɜːk/ *n.* ① [U] a type of needlework in which small pieces of cloth of different colours or designs are sewn together（不同图案杂色布块的）拼缝物；拼布工艺 ② [sing.] a thing that is made up of many different pieces or parts 拼凑之物

juxtapose /ˌdʒʌkstəˈpəʊz/	*v.* ~ **A and/with B** (*formal*) to put people or things together, especially in order to show a contrast or a new relationship between them（尤指为对比或表明其关系而）把……并置，把……并列
appeal to	to spark one's interest or appreciation 对……有吸引力
intend to	to have sth in your mind as a plan or purpose 打算；计划；想要
sit comfortably/ealily/well, etc (with sth)	to seem right, natural, suitable, etc. in a particular place or situation（在某位置或某场合）显得合适，显得自然
be hesitant to do sth	slow to speak or act because you feel uncertain, embarrassed or unwilling 犹豫的；踌躇的；不情愿的
to the point of (doing) sth	to a degree that can be described as sth 达到某种程度；近乎
link with	make, form, or suggest a connection with or between sb or sth 联系，建立连接
turn out	prove to be the case 结果（是），证明（是）
take on	to begin to have a particular quality, appearance, etc. 呈现，具有（特征、外观等）
benefit from	to be in a better position because of sth 得益于；得利于
regardless of	without being influenced by any other events or conditions 不管；无论
specialise in	to become an expert in a particular area of work, study or business; to spend more time on one area of work, etc. than on others 专门研究，专攻；专门从事

derive from ① to come or develop from sth 从……衍生出；起源于；来自 ② to get sth from sth else（从……中）得到，获得

be composed of to be made or formed from several parts, things or people 由……组成（或构成）

Reading Comprehension

Understanding the text

Answer the following questions.

1. Why did Lucienne Day have to work as a teacher?
2. What actions did the British government take to promote the nation's industrial production? And how did the couple react to this initiative?
3. Why were Day's early designs popular among consumers?
4. Why did Heals refuse to accept Day's new textile design "Calyx" at the beginning?
5. What was Lucienne Day's inspiration for the new design "Calyx" and what spirit did it embody?
6. What was her degree show? And which theme provided inspiration and innovation for Day's design?
7. Apart from furnishing fabrics, what other applications did Day produce?
8. After learning the text, what could we portray Lucienne Day as?

Critical thinking

Work in pairs and discuss the following questions.

1. In your opinion, what qualities and skills should a successful designer possess? And why?
2. How do you interpret the wisdom of the famous quotation "Every great design begins with an even better story"?
3. As a textile or fashion student, what will you do to lead you toward a bright and successful design future?

Mind mapping

The history of textile is a history of people. During the past centuries, it was the artists and designers that laid the foundations of the modern textile industry we know today. Review famous textile designers online and then present your findings with a mind map.

Language Enhancement

Words in use

Fill in the blanks with the words given below. Change the form when necessary. Each word can be used only once.

inspire	transform	initially	secure	coincide
hover	anticipate	far-reaching	encapsulate	initiative

1. He played trumpet ________ , but switched to the piano when he was in his twenties.
2. The 1964 Civil Rights Act was the most ________ anti-discrimination legislation passed by the United States Congress.
3. The demonstration is set for Sunday to ________ with World AIDS Day.
4. The UN called on all parties in the conflict to take a positive stance towards the new peace ________ .
5. He was disappointed by his failure to ________ the top job with the bank.
6. He drew much ________ from art produced by children and by primitive cultures.
7. She ________ the stereotyped image that the British have of Americans.
8. Whenever a camera was pointed at her, Marilyn would instantly ________ herself into a radiant star.
9. I could sense him behind me, ________ and building up the courage to ask me a question.

10. We've tried to ________ the most likely problems, but it's impossible to be prepared for every eventuality.

Banked cloze

Fill in the blanks by selecting suitable words from the word bank. You may not use any of the words more than once.

A. commission	F. scrutiny	K. boost
B. graphic	G. supplement	L. upholstery
C. Calyx	H. botanical	M. fabric
D. steadfast	I. Initially	N. coincide
E. juxtapose	J. secure	O. initiative

Lucienne Day was the most influential British textile designer of the 1950s. She graduated from the Printed Textiles Department at the Royal College of Art in 1940. However, it was hard to make ends meet as 1. ________ designer during World War II, and Day ended up taking teaching positions to 2. ________ her income.

At the end of the war, the British government was eager to 3. ________ its industrial production and output—especially for designers. 4. ________, Lucienne Day started out designing fabric for dresses, but she found working in the fashion industry not to her taste. A 1948 5. ________ would prove to be her big break, allowing her to leap from fashion fabrics to 6. ________ . Her designs caught the attention of powerful home furnishings company Heals, and sent her career into high orbit.

The patterns Day designed in the 1950s tended to be highly 7. ________ and energetic. On the surface they appear simple, but 8. ________ reveals that they are made up of layers of different patterns, designed to be practical and appeal both close up and at a far distance. Day's 1951 9. ________ fabric would become one of her best-known works. It was designed after 10. ________ fabrics, but it was far more stylised and abstracted. Heals resisted its use at first, but once they acquiesced they ended up selling huge quantities of it through their own company.

Expressions in use

Fill in the blanks with the expressions given below. Change the form when necessary. Each expression can be used only once.

intend to	specialise in	derive ... from	appeal to
to the point of	be composed o	regardless of	turn out

1. The committee ____________ MPs (Members of Parliament), doctors, academics and members of the public.
2. All the children are lumped together in one class, ____________ their ability.
3. Later he ____________ war photography for magazines such as *Life, Time*, and *Newsweek,* winning a number of awards.
4. He ____________ an enormous amount of satisfaction ____________ restoring old houses.
5. After all that media attention, the whole event ____________ to be a bit of a damp squib, with very few people attending.
6. His music managed to ____________ the tastes of both young and old.
7. It will notify us that you ____________ transfer a donation to our account.
8. I was an introverted child ____________ communicating with coloured crayons.

Translation

I. Translate the following paragraph into Chinese.

Raoul Dufy was a French artist who came to prominence in textile design through his collaboration with famed French fashion designer Paul Poiret. Dufy, who began his career as a painter, used bold colour and forms. Eventually, he ventured into textile design and created striking large patterns for silks, dress fabrics and decorative textiles. Attentive to detail, he studied fabric printing processes to understand dyes and translated his woodcut skills into beautiful bold patterns. His patterns were characterised by bold outlines and areas of high colour contrast. In vibrant colours, he echoed the natural world, including forms like leaves and animals like horses. Through the late 1910s and 1920s, Dufy's textile designs proved very popular. They fit perfectly into the Art Deco style of the time, which celebrated the machine age and featured geometric shapes and stylised figures. Art Deco was then the prevailing design style in Europe and the United States.

II. Translate the following paragraph into English.

黄道婆是元代一位受人尊敬的棉纺织技术改革家。她出身贫苦，幼时为童养媳，因不堪夫家虐待流落崖州（今海南岛），在此向黎族妇女学习当地的棉纺织技艺。元代元贞年间，她返回故乡，将其所学的先进纺织技术教给乡人。同时，她还不断改进纺织工具，如搅车、椎弓、三锭脚踏纺车，大大提高了纺纱效率。在织布方面，她使用错纱、配色、综线技术，织成了各种花纹棉织品，推动了松江一带棉纺织业的发展。

Paragraph Writing

How to develop a paragraph—Exemplification

A well-developed paragraph not only includes a well-focused topic sentence, but provides adequate and concrete supporting details that work together to back up the main point expressed in the topic sentence. Exemplification is a powerful rhetorical strategy to develop a paragraph and make your writing more interesting and more convincing.

An exemplification paragraph uses one or more specific, vivid and appropriate examples for the purpose of adding more information to explain, persuade, define or elaborate the main statement—the topic sentence of a paragraph. Here is a sample exemplification paragraph.

Even when a first date is a disaster, a couple can still become good friends.	→ Topic sentence
Four example, my first date with Greg was terrible. I thought he was coming to pick me up at 6:30, but instead he came at 6:00. I didn't have time top fix my hair, and my make-up looked sloppy. When I got into his car, I scraped my leg against the car door and tore my stocking.	→ Example 1
Next, he took me to an Italian restaurant for dinner, and I accidentally dropped some spaghetti on my shirt.	→ Example 2
Then we want to movie. Greg asked me which movie I wanted to see, and I chose a romantic comedy, He fell asleep during the movie, and I got angry.	→ Example 3
Now the Greg and I are good friends, we can look back and laugh at how terrible that first date was.	→ Concluding sentence

When providing examples, the following essential tips are expected to keep in mind:

1. Examples should be relevant and persuasive details that your audience can understand and believe. They can cover many different types, such as facts, statistics, quotations, personal experiences, and interviews, all of which you have seen throughout your life. They also occur in various forms of communication, whether this be an academic essay, a speech, a casual conversation, or an advertisement.
2. Examples can be organised chronologically, spatially, from the simple to complex, or with the emphatic order which moves from the first example to the one that is most important. Examples organised chronologically are moving through time, while examples organised spatially are moving through space.

In addition, to write an effective exemplification paragraph, you can use transitional words or phrases within your paragraph to create a nice bridge between sentences. Words and phrases "such as," "for example," "for instance," "to illustrate," "as an illustration," "to give a clear example," "namely" and "take …

for example" are commonly used for transition to an example.

Read the following paragraphs and locate the topic sentences, examples and concluding sentences.

1. Modern medicine focuses on illness. If a patient with a cough visits a modern doctor, then the doctor will give the patient medicine to stop the cough. If the patient also has a fever, the doctor may give different medicine to stop the fever. For every person with a cough, the doctor will probably recommend the same cough medicine. The philosophy of the modern medicine is to stop problems like coughing and fever as quickly as possible. Doctors usually see illnesses as an enemy. They use medicines like weapons to fight diseases.
2. Perhaps one of the mightiest driving forces responsible for creating trends is none other than celebrities. This is because they are already very famous and are followed by millions who like to imitate everything that they do. Let's take the example of the American reality TV star and entrepreneur, Kylie Jenner. Every time she shares a picture of herself on the photo and video sharing app, Instagram, her fans quickly go wild and try to dress the same way she dresses. There are also several fast fashion companies such as Fashion Nova that designs and models some of its outfits based on Jenner's style and markets them as a cheaper alternative to what she was originally wearing.

Write an exemplification paragraph based on the given topic sentences.

1. College degrees still exert significant influence in today's society.
2. William Morris was a driven polymath.
3. Online education provides much convenience for students.

Text B

Karl Lagerfeld: Chanel's Iconic Fashion Designer

1 When Karl Lagerfeld was appointed to Chanel in 1983, its founder Gabrielle "Coco" Chanel had been dead for 12 years, the house was **moribund** and its fashion had become a dusty shadow of its former glory. Under his **stewardship**, he revived the house codes and **rehabilitated** the business, making it one of the best-known brands in the world.

2 Creative head at Chanel for more than 30 years, Fendi since 1965 and his own labels since 1984, Lagerfeld helped **propel** Chanel to **revenues** of $9.6bn in 2017, while being ever mindful of its history.

3 "Today, not only have I lost a friend, but we have all lost an extraordinary creative mind to whom I gave carte blanche in the early 1980s to reinvent the brand," said Alain Wertheimer, the Chanel chief executive, in a statement after Lagerfeld's death.

4 Bernard Arnault, chairman and chief executive of Fendi owner LVMH, praised Lagerfeld as having the most exceptional taste and talent he has known: "I will always remember his **immense** imagination, his ability to conceive new trends for every season, his **inexhaustible** energy, the **virtuosity** of his drawings, his carefully guarded independence, his **encyclopedic** culture, his unique wit and **eloquence**."

5 Born in Hamburg in 1933, the son of Otto, a German businessman, and Elisabeth Bahlmann, a former **lingerie** saleswoman, Lagerfeld's birth date of September 10 has never been **confirmed**. It contributed in no small part to the mythology around him, and one he loved to **cultivate**. Throughout his life, he **abhorred nostalgia**.

6 "I have no notion of home," said the designer who lived in Paris for more than 50 years. His move to the French capital was encouraged by his mother who told the young Karl to get far away from his home city. His father was a businessman while he was not ever going to be in that profession. "My mother told me that Hamburg was not for me. She said: 'It's just a door—now get out of here.' And so I did."

7 In 1955, Lagerfeld was hired as Pierre Balmain's assistant after winning the coats category in a design competition now known as the Woolmark Prize the previous year. He claimed his award alongside the 18-year-old Yves Saint Laurent, who won the more **prestigious** prize in the dress category that same year. The competition established a fierce **rivalry**—and close friendship—that existed between the two designers throughout their careers, and was arguably only **exorcised** by Lagerfeld on Saint Laurent's death in 2008.

8 Although the designer established a namesake line, Roland Karl, in 1958, and worked for Tiziani, a Roman **couture** house, it was for his work with Chloé that he became first known. Lagerfeld started freelancing for the French house's **charismatic** retailer Gaby Aghion in 1964, and was later charged with **overseeing** the entire Chloé line, a **metropolitan** collection of women's clothes that embraced the **bohemian** spirit of the times. His decades-long association with the grand Italian **furrier** Fendi—he worked with all five Fendi sisters—started in 1965.

9 Lagerfeld was not a stand-and-drape designer, as he would often make clear: at the start of the season, he would present the **atelier** with **sketches** who would interpret his ideas. He had a fascination with silhouettes and innovative **fabrications**. In meetings, he would go to great lengths to emphasise the importance of a cut. He was also an accomplished caricaturist: his drawings of passers-by, from his studio window in Paris, offer a rare **testimony** to the changing texture of the times.

10 When Lagerfeld joined Chanel, he transformed the house classics into

icons: the 2.55 bag, the ballet **pump** and the **fabled tweed blazer**, all existed previously, but with each season, Lagerfeld would **tweak** and change them, enriching the Coco **legacy** and **nurturing** its growth.

11 Blessed with a genius for marketing, he directed and, frequently, photographed, the advertising campaigns. He also had a showman's sense of one-upmanship and humour. His vast, ambitious sets, such as a space launch (with rocket), a street demonstration, Mediterranean garden and a casino, ensured his show was the only ticket in town. Even in his 80s, his role in **amassing** the house's latter-day fortunes was **remarkable**.

12 Lagerfeld, who had no longtime partner, and made a point of being too busy for relationships, filled his life with fashion. An **autodidact**, his cultural insights, reading matter and interests were **voracious**.

13 An obsessive follower of **fluctuation** in the zeitgeist, he once **professed** to owning four iPhones and 20 to 30 iPads. But he was **stubborn** about other technologies: email was **anathema**, and requests to his studio were still made by fax. In interviews he had a **conspiratorial** nature, and a sense of **devilry**, often dispensing with press aides and brand representatives in order to speak more **candidly** one-to-one. Of the dozens of collaborators who worked for him, all remarked on his **brilliant** sense of fun.

14 He also **indulged** extraordinary obsessions: for the past decade he was synonymous with his high-necked white collar and **slimline** suits; for a year he drank little other than Diet Coke (for which he designed a bottle). But few things were so dear to him as his cat Choupette, on whom he doted to an almost Marie Antoinette-ish degree, using several nannies to tend it, and casting its likeness in numerous jewels.

15 Lagerfeld's influence cannot be overestimated. At Chanel and Fendi, he enjoyed an extraordinary vantage point, floating seamlessly between the different groups. In addition to designing, he launched careers—Lily Allen, Vanessa Paradis, Ines de la Fressange, Kate Moss, among them—shaped designers, inspired editors, **remoulded** the style of the red carpet and published dozens of books. His absence at the Chanel couture show in January, at which he asked his creative studio director Virginie Viard, to take the final bow, was notable in being the only finale he had ever missed.

16 Lagerfeld's closest collaborator for more than 30 years, Viard has been **entrusted** with the creative work for the collections, so that the legacy of Gabrielle Chanel and Karl Lagerfeld can live on. The world lost a giant, but his genius touched the lives of so many people and they will never forget his incredible talent and endless inspiration.

Notes

The International Woolmark Prize The award celebrates outstanding fashion talents from around the globe who showcase the beauty and versatility of Merino wool.

Gabrielle Coco Chanel 加布里埃·可可·香奈儿

Alain Wertheimer 阿兰·韦特海默（奢侈品牌香奈儿以及化妆品牌 Bourjois 的拥有人）

Bernard Arnault 伯纳德·阿尔诺 (LVMH 的董事长兼首席执行官)

Hamburg 汉堡（德国城市）

Marie Antoinette 玛丽·安托瓦内特（法王路易十六的王后）；绝代艳后

New words and phrases

moribund /ˈmɒrɪbʌnd/ *a.* ① (especially of an organisation or business) not active or successful and may be coming to an end (尤指组织或公司)无生气的; 停滞不前的; 即将倒闭的 ② in a very bad condition; dying 垂死的；濒临死亡的；奄奄一息的

stewardship /ˈstjuːədʃɪp/ *n.* [U] the way of taking care of or managing sth, for example property, an organisation, money or valuable objects 管理方法；组织方式

rehabilitate /ˌriːəˈbɪlɪteɪt/ *v.* ① to return sth to its previous good condition 使恢复原状；修复 ② to return someone to a good, healthy, or normal life or condition after they have been in prison, been very ill, etc.（监禁或病后）使康复；使恢复正常生活

propel /prəˈpel/ *v.* ① to move, drive or push sth forward or in a particular direction 推动；驱动；推进 ② to force sb to move in a particular direction or to get into a particular situation 驱使；迫使；推搡

revenue /ˈrevənjuː/ *n.* [U]（also revenues [pl.]）the money that a government receives from taxes or that an organisation, etc. receives from its business（政府的）税收，岁入；（公司的）收益

immense /ɪˈmens/ *a.* extremely large or great, enormous 极大的；巨大的

inexhaustible /ˌɪnɪɡˈzɔːstəbl/ *a.* (of an amount or supply of sth) unable to be used up because existing in abundance（某物的数量或供应）取之不尽的，用之不竭的

virtuosity /ˌvɜːtʃuˈɒsəti/ *n.* [U] the quality of being extremely skilled at something 高超技艺

encyclopedic /ɪnˌsaɪkləˈpiːdɪk/ *a.* covering a large range of knowledge, often in great detail 百科全书似的；包罗万象的；博学的

eloquence /ˈeləkwəns/ *n.* [U] fluent or persuasive speaking or writing 雄辩，流利的口才；（讲话或文字的）说服力，生动流畅

lingerie /ˈlænʒəri/ *n.* [U] women's underwear 女内衣

confirm /kən'fɜːm/ — *v.* ① to state or show that sth is definitely true or correct, especially by providing evidence（尤指提供证据来）证实，证明，确认 ② to make an arrangement or meeting certain 确认，确定（安排或会议）

cultivate /'kʌltɪveɪt/ — *v.* ① to develop an attitude, image, or skill and make it stronger or better 培养(态度、技巧等); 树立(形象、观念等) ② to prepare land and grow crops on it, or to grow a particular crop 耕作；栽培；种植

abhor /əb'hɔː(r)/ — *v.* to hate a way of behaving or thinking, often because you think it is not moral 憎恶；憎恨；厌恶

nostalgia /nɒ'stældʒə/ — *n.* [U] a feeling of pleasure and also slight sadness when you think about things that happened in the past 对往事的怀念，怀旧，念旧

prestigious /pre'stɪdʒəs/ — *a.* very much respected and admired, usually because of being important 有威望的，有声望的，受尊敬的

fierce /fɪəs/ — *a.* ① showing strong feelings or energetic activity 强烈的；狂热的 ② (people or animals) angry and aggressive in a way that is frightening 凶猛的；凶狠的；凶残的

rivalry /'raɪvlri/ — *n.* [C, U] a state in which two people, companies, etc. are competing for the same thing 竞争；竞赛；较量

exorcise /ˈeksɔːsaɪz/ *v.* ① to remove the bad effects of a frightening or upsetting event 消除，除去，忘掉（可怕或烦恼的事情产生的后果）② to force an evil spirit to leave a person or place by using prayers or magic （通过祈祷或魔法）给……驱除邪魔

couture /kuˈtjʊə(r)/ *n.* [U] the designing and making of expensive fashionable clothes, or the clothes themselves 高级时装设计制作；高级时装

charismatic /ˌkærɪzˈmætɪk/ *a.* ① A charismatic person attracts, influences, and inspires people by their personal qualities. 有魅力的；有号召力（或感召力）的 ② believing in special gifts from God; worshipping in a very enthusiastic way 蒙受神恩的；有特恩的；虔诚崇拜的

oversee /ˌəʊvəˈsiː/ *v.* to watch or organise a job or an activity to make certain that it is being done correctly 监督；监察；监管

metropolitan /ˌmetrəˈpɒlɪtən/ *a.* ① connected with a large or capital city 大城市的；大都会的 ② connected with a particular country rather than with the other regions of the world that the country controls 本土的

bohemian /bəʊˈhiːmiən/ *a.* to describe artistic people who live in an unconventional way 不受世俗陈规束缚的
n. [C] a person, often sb who is involved with the arts, who lives in a very informal way without following accepted rules of behaviour 行为举止不墨守陈规者；放荡不羁的艺术家

furrier /ˈfʌrɪə(r)/ *n.* [C] a person who prepares or sells clothes made from fur 毛皮加工者；皮货商

atelier /əˈteliei/ *n*. [C] a room or building in which an artist works（艺术家的）工作室，制作室

sketch /sketʃ/ *n*. [C] A sketch is a drawing that is done quickly without a lot of details. Artists often use sketches as a preparation for a more detailed painting or drawing. 草图；略图；素描

v. ① If you sketch something, you make a quick, rough drawing of it. 画……的素描；画……的速写 ② If you sketch a situation or incident, you give a short description of it, including only the most important facts. 概述；简述

fabrication /ˌfæbrɪˈkeɪʃn/ *n*. [U] the act or process of making or producing goods, equipment, etc. from various different materials 制造；装配；组装

caricaturist /ˈkærɪkəˌtʃʊəist/ *n*. [C] a person who shows other people in an exaggerated way in order to be humorous or critical, especially in drawings or cartoons 漫画家

testimony /ˈtestɪməni/ *n*. ① [U, sing.] a thing that shows that sth else exists or is true 证据；证明 ② [C, U] a formal written or spoken statement saying what you know to be true, usually in court 证词；证言；口供

pump /pʌmp/ *n*. [C] ① a light soft shoe that you wear for dancing or exercise 轻软舞鞋；轻便帆布鞋 ② a machine that is used to force liquid, gas or air into or out of sth 抽水机；泵；打气筒

v. to make water, air, gas, etc. flow in a particular direction by using a pump or sth that works like a pump 用泵（或泵样器官等）输送

fabled /ˈfeɪbld/ *a.* famous and often talked about, but rarely seen 传说中的；著名的

tweed /twiːd/ *n.* [U] a thick woolen cloth, often woven from different coloured threads 粗花呢

blazer /ˈbleɪzə(r)/ *n.* [C] a jacket, not worn with matching trousers/ pants, often showing the colours or badge of a club, school, team, etc.（常带有俱乐部、学校、运动队等的颜色或徽章的）夹克

tweak /twiːk/ *v.* ① improve sth such as a system or a design by making a slight change 稍稍改进（如系统或设计）② to pull or twist sth suddenly 扭；拧；扯

legacy /ˈlegəsi/ *n.* [C] ① money or property that is given to you by sb when they die 遗产；遗赠财物 ② sth that is a part of the history or that remains from an earlier time 历史遗产，遗留物

nurture /ˈnɜːtʃə(r)/ *v.* ① to help sb/sth to develop and be successful 扶持；帮助；支持；促进 ② to care for and protect sb/sth while they are growing and developing 养育；养护；培养

amass /əˈmæs/ *v.* to get a large amount of sth, especially money or information, by collecting it over a long period 集聚，积累，大量收集（尤指钱或信息）

remarkable /rɪˈmɑːkəbl/ *a.* unusual or special and therefore surprising and worth mentioning 非凡的；奇异的；引人注目的

autodidact /ˈɔːtəʊdɪdækt/ *n.* [C] a person who has taught himself or herself sth rather than having lessons 自学者；自修者

voracious /vəˈreɪʃəs/

a. ① wanting a lot of new information and knowledge（对信息、知识）渴求的；求知欲强的；如饥似渴的 ② eating or wanting large amounts of food 饭量大的；贪吃的；狼吞虎咽的

fluctuation /ˌflʌktʃuˈeɪʃn/

n. [C, U] a change, or the process of changing, especially continuously between one level or thing and another 波动；涨落；起伏

profess /prəˈfes/

v. ① to state openly that you have a particular belief, feeling, etc. 宣称；公开表明 ② to claim that sth is true or correct, especially when it is not 声称；自称；谎称

stubborn /ˈstʌbən/

a. ① determined not to change your opinion or attitude 固执的；执拗的；顽固的；倔强的 ② difficult to get rid of or deal with 难以去除（或对付）的

anathema /əˈnæθəmə/

n. [C, usually sing.] a thing or an idea which you hate because it is the opposite of what you believe 可憎的事物；可恶的想法

conspiratorial /kənˌspɪrəˈtɔːriəl/

a. ① relating to a secret plan to do sth bad, illegal, or against someone's wishes 阴谋的；密谋的；似阴谋的 ② showing that you share a secret 会意的；心照不宣的

devilry /ˈdev(ə)lrɪ/

n. [U] reckless or malicious fun or mischief 鲁莽的、恶意的胡闹

candidly /ˈkændɪdli/

ad. in an honest and direct way that people might not like 坦率地；直率地

brilliant /ˈbrɪliənt/	*a.* ① very good 非常好的，出色的 ② extremely intelligent or skillful 颇有才气的，聪颖的；技艺高超的 ③ (of light or colours) very bright （光线或色彩）明亮的；鲜艳的
indulge /ɪnˈdʌldʒ/	*v.* ① to allow yourself to have or do sth that you like, especially sth that is considered bad for you 沉湎，沉迷，沉溺于…… ② to be too generous in allowing sb to have or do whatever they like 放纵；听任
slimline /ˈslɪmlaɪn/	*a.* smaller or thinner in design than usual 式样小巧的；薄型的
remould /ˌriːˈməʊld/	*v.* change or refashion the appearance, structure, or character of 翻新；改造；重新塑造
entrust /ɪnˈtrʌst/	*v.* ~ **A (to B)** | ~ **B with A** to make sb responsible for doing sth or taking care of sb 委托；交托；托付
appoint to	to choose someone for a position or a job 委派任……职位，任命
under one's stewardship	the way in which an organisation or event is controlled by someone 在某人的管理下
mindful of sb/sth	remembering sb/sth and considering them or it when you do sth 记着；想着；考虑到
in no small part	to an important degree 很大程度上，相当一部分
charge sb with sth	① to entrust (someone) with a task as a duty or responsibility 使承担（职责或责任） ② accuse sb of sth, especially an offence under law 指责（某人）；控告（某人）

go to great lengths to do sth	to put a lot of effort into doing sth, especially when this seems extreme 竭尽全力；不遗余力
fascination with/for sb/sth	the state of being very attracted to and interested in sb/sth 入迷；着迷
be blessed with sth	to be lucky in having a particular good quality or skill 幸运地拥有 ……
make a point of doing sth	to take paticular care to do something or always do something 特意做某事，总是要做某事；一贯注重；重视
dispense with sth/sb	to get rid of or stop using sth or sb that you do not need 免除；废止；省去；不再使用
remark on	to notice sth and make a remark about it 谈论，评论（觉察到的事）
be synonymous with sth	be closely associated with or suggestive of sth 密不可分的；是 …… 的代名词
other than	(usually used in negative sentences) except（通常用于否定句）除 …… 以外
dote on/upon	to love or care about sb/sth very much and ignore any faults they may have 溺爱
take a bow	(of an actor or entertainer) acknowledge applause after a performance by bowing 鞠躬答礼，答谢；谢幕
live on	to continue to live or exist 继续活着；继续存在

Reading Comprehension

Understanding the text

Choose the best answer to each of the following questions.

1. Under what circumstance was Karl Lagerfeld appointed to Channel?

   A. As soon as he graduated from school.

   B. When the house was moribund.

   C. Immediately after Gabrielle Coco Chanel died.

   D. When the house was flourishing.

2. According to Bernard Arnault, Lagerfeld has the most unusual taste and talent for the following except ________.

   A. his immense imagination

   B. his ability to conceive new trends for every season

   C. his stewardship

   D. his encyclopedic culture and unique wit

3. When did Karl Lagerfeld first win the Woolmark Prize?

   A. In 1954.

   B. In 1955.

   C. In 1958.

   D. In 1965.

4. Which of the following statements about Karl Lagerfeld is false in Paragraphs 11–12?

   A. He frequently directed and photographed the advertising campaigns.

   B. His vast ambitious sets ensured that his show was the only ticket in town.

   C. He was busy in building relationships.

   D. His cultural insights, reading matter and interests were voracious.

5. After Lagerfeld's death, who has been entrusted with the creative work so that the legacy of Gabrielle Chanel can live on?

   A. Alain Wertheimer.

   B. Bernard Arnault.

   C. Pierre Balmain.

   D. Virginie Viard.

Research project

This passage introduces Karl Lagerfeld's legendary life on the basis of other people's comments. Please search for more information about his life story and fashion career to present him in a panoramic view.

Language Enhancement

Words in use

Fill in the blanks with the words given below. Change the form when necessary. Each word can be used only once.

amass	metropolitan	nostalgia	moribund	fierce
revenue	fluctuation	profess	nurture	immense

1. This serves as an introduction to which languages are safe, which are endangered, and which are ________.
2. Television companies have been massaging their viewing figures in order to attract more advertising ________.
3. He ________ ignorance of the whole affair, though I'm not sure I believe him.
4. They spent an ________ amount of time getting the engine into perfect condition.
5. He was drawn to the ________ glamour and excitement of Paris.
6. Some of his colleagues envy the enormous wealth that he has ________.
7. Employers can adjust their workforce in line with ________ in demand for goods and services.
8. The expansion plans will face ________ resistance from environmentalists.
9. As I drove into the city, I felt a wave of ________ sweep over me.
10. As a record company executive, his job is to ________ young talent.

Expressions in use

Fill in the blanks with the expressions given below. Change the form when necessary. Each expression can be used only once.

appoint to	dispense with	charge with
under the stewardship of	be blessed with	mindful of
remark on	make a point of	

1. He said that the firm, ____________ his assistants, would suffer no disruptions.
2. ____________ the poor road conditions, she reduced her speed to 30 mph.
3. I ____________ the recent profusion of books and articles on the matter.
4. Several England supporters were arrested and ____________ disturbing the peace after the match.
5. She ____________ keeping all her shopping receipts.
6. Princess Dianna ____________ both beauty and brains.
7. She has had to ____________ a lot of luxuries since her husband lost his job.
8. The truth is that they ____________ no more than a token number of women ____________ managerial jobs.

Sentence structure

I. Complete the following sentences by translating the Chinese into English, using "cannot be overestimated/overstated/overemphasised" structure.

Model: Karl Lagerfeld's influence ________________________ (无论怎样高估也不为过).

→ Karl Lagerfeld's influence cannot be overestimated.

1. During the pandemic, the need for wearing of face masks while on public transportation ____ ____________________________________ (无论怎样强调也不为过).

2. The importance of strict hygiene in the preparation of food ________________

____（无论怎样夸大也不为过）.

3. The influence of early education on child growth and development __________

____（无论怎样高估也不为过）.

II. Combine the following sentences by using "while" to indicate a contrast. Make changes when necessary.

Model: His father was a businessman. He was not ever going to be in that profession.

→ His father was a businessman while he was not ever going to be in that profession.

1. Tom is very extrovert and confident. Katy is shy and quiet.

____

____.

2. I prefer working late into the night. He would rather work early before dawn.

____

____.

3. Schools in the north tend to be better equipped. Schools in the south are relatively poor.

____

____.

Extensive Reading

Chinese Designer Uma Wang: Creating Costumes for Film Sets

1 Talented and commercially successful designer Uma Wang may be celebrating her eponymous brand's 10th anniversary this year, but when

the chance came to create costumes for a historical movie, nerves began to flutter and doubts set in.

2 The fears were unfounded. Wang's silver screen debut, designing costumes for the 1930s Beijing-set drama *Hidden Man*, directed by Jiang Wen and starring Liao Fan and Eddie Peng, was deemed an unqualified success. The film even got a nomination for best make-up and costume design.

3 Wang found the process creatively stimulating and professionally satisfying, not to mention immensely challenging, necessitating a total change of mindset from fashion design.

4 "When I design my collections, I don't know who will wear it. But when you do a movie, you know exactly who will wear it, so there is a real connection—you know the personality and the story, so you have to inject feeling and emotion into the mood of the person," Wang says. "The director said he wanted someone with no experience, like a blank sheet of paper, so I would think in a different way. Sometimes experience can kill the magic of creativity."

5 Wang designed the costumes for the lead actor and actress and explains that the period was the "end of an era".

6 "It was a traditional design, a big volume with big shoulders and dramatic proportions. You can still see the influence from the Qing era. I had to do a lot of research. Beijing is very cold in winter so you need layers, which was interesting. I made 15 different looks in total."

7 Now she's working on costumes for a Hong Kong movie set in contemporary times with a dark, drug-related theme, directed by Juno Mak. She explains that the assignment is all about details and subtle shading, as the main characters dress mostly in black suits.

8 Wang's extra movie gigs mean that an already hectic travel schedule is even more frantic. Unusually, the brand has design offices in Shanghai

and a manufacturing and distribution base in Italy, with Wang shuttling between the two regularly.

9 During Italian sojourns, home is a villa near Verona, the city where Shakespeare's *Romeo and Juliet* was set. Located in a small village, the villa is conveniently close to her brand's factory, making it easy for her to liaise with local staff and suppliers.

10 Wang is one of only a handful of truly international Chinese designers whose work can be found in the big-name stores of Paris, Milan, London and New York. Her clothes are in around 120 outlets around the world, in 40 different countries, retailing for an average of USD 650 for a dress or jacket and USD 1,300 for a coat.

11 11The beginnings of the Uma Wang brand were modest. The first significant sales were in Moscow, followed by interest in London and Milan. Wang, a graduate of London's Central Saint Martins fashion school, maintains that the quality of the Italian-produced fabric is a lure to fashionistas, as is the finish, which results in a worn-in vintage feel.

12 "I produce the clothes in Italy not just because of the quality but because of the way the Italians work. They try to understand what is behind my style and philosophy," she says.

13 Wang, who in the past has shown her collections at Milan and Paris fashion weeks, is among a handful of Chinese designers who have cancelled their plans to hold shows during Paris Fashion Week, which begins on February 24.

14 She explains her customers have "special tastes" and many are artists, dancers, people working in the film business and generally international people.

15 "They love Uma Wang, the whole aesthetic, not just the fabric. The style is long and flowing. For me the garment is not to show too much the shape of the woman; it is more hiding with the big volume—this is the philosophy of the East. It is typical of Chinese philosophy, although I don't use so many elements from China in the clothes, but customers see that influence from the shape and the cut."

16 Until her late teens, Wang was known more for her basketball talent than fashion sense. The native of China's Hebei Province attended one of the nation's specialist sports schools from the age of 12, thriving in the fiercely competitive environment.

17 During end-of-term theatrical shows, where students made their own costumes, a talent for design and the ability to make best use of whatever materials were to hand began to emerge. The next step in her roundabout route to fashion was spotting a newspaper cutting that referenced a design course at a textile university. Until then, Wang was unaware such institutions existed.

18 The swap from shooting hoops to creating patterns led to further schooling in China, education at Central Saint Martins, and then the launch of Uma Wang 10 years ago. A career boost came when she was awarded a place on the CFDA/Vogue Fashion Fund's China Exchange Programme, which saw her receive six weeks of training in New York.

19 Warm, gently spoken and immediately likeable, Wang is clearly happy with her lot in life. She constantly expresses surprise, with one of her trademark grins, that so many lucky breaks have come her way.

20 Hers has been a varied career, which keeps taking new twists and turns. As well as the movie gigs, Wang is looking to expand into homeware, encouraged by customers who are convinced her aesthetic would cross over well.

21 But after hearing about Wang's early life, one question begs to be asked: does she still cut the mustard on the basketball court?

22 You bet she can, aided and abetted by a pair of extra-large hands and long fingers, which she proudly displays, explaining how they make it easier to grasp the ball.

23 "If I finish work early when I am in Shanghai I play with the young kids in the park," she says. "I am still pretty fit. At the beginning they called me Uma Auntie. When they realized I can play, they changed to it Uma Sister."

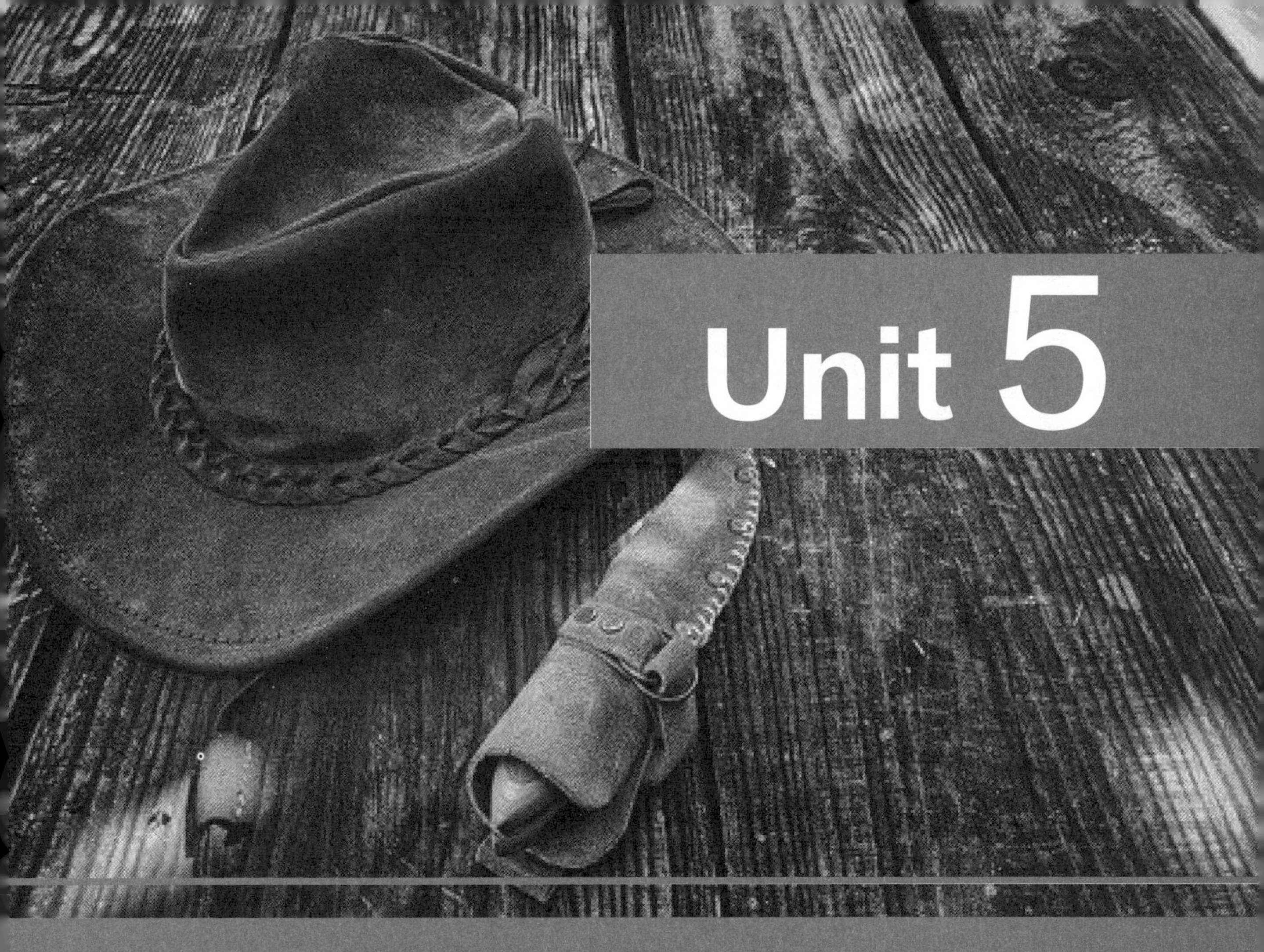

Unit 5

Western Costume and Culture

Clothes make a statement. Costumes tell a story.

— *Mason Cooley*

Most people dress according to their culture and their ethnic background. You almost embrace the culture, and now it's taken away. That's tough.

— *Dale Davis*

Pre-Reading Activities

1. Watch the video and choose the best answer to each of the following questions.

(1) Why do "the patterns of the eyes and the ears" appear everywhere on Queen Elizabeth's dress?

A. Because she liked those patterns.

B. Because the tailor designed the dress like that.

C. Because her rule of stage is symbolised by eyes and ears.

(2) What's the meaning of the rainbow on Queen Elizabeth's portrait?

A. An element used to make the colour of the portrait more gorgeous.

B. An element used as a metaphor and symbol of peace.

C. An element used to show Queen Elizabeth's amiability.

2. Watch the video again and fill in the blanks.

(1) Whilst Elizabeth's incredible dress may have been ______________, it certainly gave her________.

(2) It wasn't just the ________ that had ________, the dress is full of ________ that an Elizabethan viewer could read like a book.

(3) The pearls signified ________. The bejewelled snake symbolised ________, the heart symbolised ________.

3. Work in pairs and discuss the following questions with your partner.

(1) Have you ever noticed there is symbolism used on people's clothes in your daily life? Could you share some examples?

(2) Do the same patterns or elements used on clothes have the same meaning in different cultures? Give examples to confirm your answer.

Text A

Everything You Ever Wanted to Know about the White Wedding Dress

1 Many different cultures wear different colours on the wedding day. In China and India, for example, red dresses are worn as a symbol of good luck and success. Traditional Nigerian brides veer towards brightly coloured, **elaborately accessorized** dresses, and wedding dresses in Ghana vary from couple to couple, with each family uses its own **intricate** cloth pattern. Traditional Hungarian dresses are white with colourful **floral patterns** embroidered down their lengths, and Malaysian **gowns** usually come in purple or **violet**. However, the most widely accepted and chosen colour for wedding dresses is white. And it is Queen Victoria who lead the trend.

2 Queen Victoria set two strong fashion trends during her lifetime: deep black for **mourning** and white wedding dresses. Although black had been worn in the western world for mourning since Roman times, Queen Victoria elevated it to another level. Before her, **royal** brides wore wedding dresses in a variety of **hues**, with red being one of the most popular, while white dresses were reserved for women who were being presented at **court**.

3 Intent on making a statement, the fashion-loving Queen chose a non-traditional dress and flower **crown** for her wedding to Prince Albert on 10 February 1840, which she said was "the happiest day of my life." The dress was made from Spitalfields cream silk-satin with a **flounce** of Honiton lace at the neck and sleeves. With its slim waist, full **crinoline petticoat** and lace **embellishments**, it's still considered the "classic" wedding dress silhouette in the west today.

4 But why white? The reason is **layered**. Knowing that her dress would be reported on around the world, Queen Victoria chose to wear a dress **trimmed** with handmade Honiton lace from the small village of Beer, to support the **declining** lace trade and give the industry a **boon**. White, she **reasoned**, was the best way to show off the lace makers' **artistry**.

5 That's not to say that her marriage to Prince Albert was purely a show of duty: there were just as many **romantic** elements involved. Part of the reason Queen Victoria **eschewed** the **heirloom** jewels, heavy fabrics and rich colours was that she didn't want to make her vows to her husband as a **monarch**, but rather as the woman he loved.

6 As accounts of Victoria's wedding spread, other European leaders followed suit. The new dresses were **conspicuously luxurious**: **laundering** clothing was taxing in the 19th century and white dresses were hard to maintain. Unlike today, wedding gowns were worn several times during a lifetime; even Queen Victoria brought hers out for other events. As white dresses gained popularity for weddings, they gained new **symbolism**—the colour began to **signify** purity and innocence, in addition to wealth. White also looked good in early black-and-white or sepia-toned photography.

7 However, it would take another few decades for white wedding dresses to be **democratised** among middle-class marrieds in Europe and the US. Until then, many women simply wore their nicest dress on their wedding day. As society became more **prosperous** in the **aftermath** of the Second World War and clothing became cheaper to produce, the white, single-use wedding dress—and lavish party to show it off—became a **distinctive** part of getting married.

8 The **portrayal** of weddings in Hollywood, as well as the speed and ease in which people could see images of **celebrity** weddings, helped **cement** the notion that marriage demanded a white dress. In 1956, film footage and photographs of Grace Kelly in her wedding gown, made from lace, silk, **pearls** and **tulle**, quickly made their way across the globe. In 1981,

750 million people watched Charles, Prince of Wales marry Lady Diana Spencer in her **ivory** silk taffeta gown with 25ft train designed by David and Elizabeth Emanuel. More recently, Kate Middleton's dress by Sarah Burtan (Alexander McQueen) and Meghan Markle's dress by Clare Waight Keller for Givenchy inspired copies overnight.

9 And then there's the tradition to close couture shows with a white wedding gown. Designers have long made wedding dresses for private clients and, during the first half of the 20th century, these gowns sometimes made the shows of the summer collections. Jeanne Lanvin's white wedding dress, designed for the marriage of her beloved daughter Marguerite Marie-Blanche to the Comte Jean de Polignac in 1924, is a **poignant** example. While the idea of closing a couture show with a white wedding dress may date to the '40s or '50s, by 1957 it had become a tradition—a *Vogue*'s article from April that year states that "spring's Paris Collections ... traditionally close with the presentation of a bride's dress." Some of these showstoppers, such as Yves Saint Laurent's **cocoon** dress from 1965, have become **iconic**.

10 Today, even in cultures where white wedding dresses are not the **norm**, such as China—traditionally red symbolises luck and prosperity—some brides change into white dresses for official photographs. And although the white dress is sometimes replaced with a white trouser suit, the colour remains a top choice for celebrating a union.

11 A few more recent celebrity weddings, however, could **initiate** a break with tradition. Reese Witherspoon's blush-pink gown from her 2011 wedding increased the sales of pastel-hued wedding dresses in some of the established bridal **boutiques** in the US; while on the **runway**, Adut Akech closed the Chanel autumn/winter 2018 couture show in a mint-green two-piece tweed suit. Now, 180 years after Queen Victoria's wedding, the time may have come to bring some colour back to the big event.

Notes

Queen Victoria Queen Victoria served as monarch of Great Britain and Ireland from 1837 until her death in 1901. She became Empress of India in 1877. After Queen Elizabeth II, Victoria is the second-longest reigning British monarch. Victoria's reign saw great cultural expansion; advances in industry, science and communications; and the building of railways and the London Underground.

Grace Kelly Grace Kelly rose to fame as a leading Hollywood actress following her prominent role in *High Noon*. Along with her Academy Award-winning performance in *The Country Girl*, she starred in the Alfred Hitchcock films *Rear Window*, *Dial M for Murder* and *To Catch a Thief*. Kelly left Hollywood behind after marrying Prince Rainier III of Monaco in 1956, thereby becoming known as Princess Grace. She died in her adopted home country in 1982, following a car accident.

Ghana 加纳（非洲西部国家）

Prince Albert 阿尔伯特亲王

New words and phrases

elaborately /ɪˈlæbərətli/ *ad.* in a very complicated and detailed way; in a carefully prepared and organised way 复杂地；详尽地；精心制作地

accessorize /əkˈsesəraɪz/ *v.* to add fashionable items or extra decorations to sth, especially to your clothes 加时髦配件，加装饰物，加搭配服饰品（于服装等）

intricate /ˈɪntrɪkət/ *a.* having a lot of different parts and small details that fit together 错综复杂的

floral /ˈflɔːrəl/ *a.* ① consisting of pictures of flowers; decorated with pictures of flowers 绘有花的；饰以花的 ② made of flowers 由花组成的

pattern /ˈpætn/ *n.* [C] ① a regular arrangement of lines, shapes, colours, etc. as a design on material, carpets, etc. 图案；花样；式样 ② a design, set of instructions or shape to cut around that you use in order to make sth 模型；底样；纸样

gown /gaʊn/ *n.* [C] a woman's dress, especially a long one for special occasions（尤指特别场合穿的）女裙；女长服；女礼服

violet /ˈvaɪələt/ *n.* ① [C] a small wild or garden plant with purple or white flowers with a sweet smell that appear in spring 紫罗兰 ② [U] a bluish-purple colour 蓝紫色；紫罗兰色

mourn /mɔːn/ *v.* ~ **(for sb/sth)** to feel and show sadness because sb has died; to feel sad because sth no longer exists or is no longer the same（因失去……而）哀悼，忧伤

royal /ˈrɔɪəl/ *a.* ① connected with or belonging to the king or queen of a country 国王的；女王的；皇家的；王室的 ② impressive; suitable for a king or queen 庄严的；盛大的；高贵的；适合国王（或女王）的

hue /hjuː/ *n.* [C] (*literary*) a colour; a particular shade of a colour 颜色；色度；色调

court /kɔːt/ *n.* [C, U] ① the place where legal trials take place and where crimes, etc. are judged 法院；法庭；审判庭 ② the official place where kings and queens live 王宫；宫殿；宫廷

crown /kraʊn/ *n.* [C] an object in the shape of a circle, usually made of gold and precious stones , that a king or queen wears on his or her head on official occasions 王冠；皇冠；冕

flounce /flaʊns/	*n*. [C] a strip of cloth that is sewn around the edge of a skirt, dress, curtain, etc.（衣、裙、窗帘等的）荷叶边
crinoline /ˈkrɪnəlɪn/	*n*. [C] a frame that was worn under a skirt by some women in the past in order to give the skirt a very round full shape（旧时的）裙衬，裙撑，裙架
petticoat /ˈpetɪkəʊt/	*n*. [C] (*old-fashioned*) a piece of women's underwear like a thin dress or skirt, worn under a dress or skirt 衬裙
embellishment /ɪmˈbelɪʃmənt/	*n*. [C] a decoration added to sth to make it seem more attractive or interesting 装饰
layered /ˈleɪəd/	*a*. something that is made or exists in layers 分层的；层叠的
trim /trɪm/	*v*. ① to make sth neater, smaller, better, etc., by cutting parts from it 修剪；修整 ② to cut away unnecessary parts from sth 切去，割掉，剪下，除去（不必要的部分）③ [usually passive] ~ **sth (with sth)** to decorate sth, especially around its edges 装饰，修饰，点缀（尤指某物的边缘）
declining /dɪˈklaɪnɪŋ/	*a*. becoming lower, smaller or weaker 减少的；下降的；衰弱的；衰退的
boon /buːn/	*n*. [C] something that is very helpful and makes life easier for you 非常有用的东西；益处
reason /ˈriːzn/	*v*. ① to form a judgement about a situation by considering the facts and using your power to think in a logical way 推理；推论；推断 ② to use your power to think and understand 思考；理解

artistry /ˈɑːtɪstri/ *n.* [U] the skill of an artist 艺术技巧

romantic /rəʊˈmæntɪk/ *a.* ① connected or concerned with love or a sexual relationship 浪漫的；爱情的；情爱的 ② beautiful in a way that makes you think of love or feel strong emotions 浪漫的；富有情调的；美妙的

eschew /ɪsˈtʃuː/ *v.* (*formal*) to deliberately avoid or keep away from sth （有意地）避开，回避，避免

heirloom /ˈeəluːm/ *n.* [C] a valuable object that has belonged to the same family for many years 传家宝；世代相传之物

monarch /ˈmɒnək/ *n.* [C] a person who rules a country, for example a king or a queen 君主；帝王

conspicuously /kənˈspɪkjuəsli/ *ad.* in a way that is easy to see or notice, or that is likely to attract attention 易见地；明显地；惹人注意地

luxurious /lʌgˈʒʊəriəs/ *a.* very comfortable; containing expensive and enjoyable things 十分舒适的；奢侈的

launder /ˈlɔːndə(r)/ *v.* (*formal*) to wash, dry and iron clothes, etc. 洗熨（衣物）

popularity /ˌpɒpjuˈlærəti/ *n.* [U] ~ **(with/among sb)** the state of being liked, enjoyed or supported by a large number of people 受欢迎；普及；流行

symbolism /ˈsɪmbəlɪzəm/ *n.* [U] the use of symbols to represent ideas, especially in art and literature（尤指文艺中的）象征主义，象征手法

signify /ˈsɪgnɪfaɪ/ *v.* ① to be a sign of sth 表示；说明；预示 ② to do sth to make your feelings, intentions, etc. known 表达，表示，显示（感情、意愿等）

democratise /dɪˈmɒkrətaɪz/ — *v.* (*formal*) to make a country or an institution more democratic 使民主化

prosperous /ˈprɒspərəs/ — *a.* rich and successful 繁荣的；成功的；兴旺的

aftermath /ˈɑːftəmæθ/ — *n.* [usually sing.] the situation that exists as a result of an important (and usually unpleasant) event, especially a war, an accident, etc.（战争、事故、不快事情的）后果，创伤

distinctive /dɪˈstɪŋktɪv/ — *a.* having a quality or characteristic that makes sth different and easily noticed 独特的；特别的；有特色的

portrayal /pɔːˈtreɪəl/ — *n.* [C, U] the act of showing or describing sb/sth in a picture, play, book, etc.; a particular way in which this is done 描绘；描述；描写；展现方式

celebrity /səˈlebrəti/ — *n.* ① [C] a famous person 名人；名流 ② [U] the state of being famous 名望；名誉；著名

cement /sɪˈment/ — *v.* ① [often passive] ~ **A and B (together)** to join two things together using cement, glue, etc.（用水泥、胶等）粘结，胶合 ② to make a relationship, an agreement, etc. stronger 加强，巩固（关系等）

pearl /pɜːl/ — *n.* [C] a small hard shiny white ball that forms inside the shell of an oyster and is of great value as a jewel 珍珠

tulle /tjuːl/ — *n.* [U] a type of soft fine cloth made of silk, nylon, etc. and full of very small holes, used especially for making veils and dresses 绢网，丝网眼纱，网眼织物（尤用以制作面纱或连衣裙）

ivory /ˈaɪvəri/ — *n.* [U] a yellowish-white colour 象牙色；乳白色

poignant /ˈpɔɪnjənt/ *a.* having a strong effect on your feelings, especially in a way that makes you feel sad 令人沉痛的；悲惨的；酸楚的

cocoon /kəˈkuːn/ *n.* [C] ① a covering of silk threads that some insects make to protect themselves before they become adults 茧
② a soft covering that wraps all around a person or thing and forms a protection 保护膜；防护层；软罩

iconic /aɪˈkɒnɪk/ *a.* acting as a sign or symbol of sth 符号的；图标的；图符的；偶像的

norm /nɔːm/ *n.* ① (often **the norm**) [sing.] a situation or a pattern of behaviour that is usual or expected 常态；正常行为 ② (**norms**) [pl.] standards of behaviour that are typical of or accepted within a particular group or society 规范；行为标准 ③ [C] a required or agreed standard, amount, etc. 标准；定额；定量

initiate /ɪˈnɪʃieɪt/ *v.* (*formal*) to make sth begin 开始；发起；创始

boutique /buːˈtiːk/ *n.* [C] a small shop/store that sells fashionable clothes or expensive gifts 时装店；精品店；礼品店

runway /ˈrʌnweɪ/ *n.* [C] ① a long narrow strip of ground with a hard surface that an aircraft takes off from and lands on 飞机跑道 ② the long stage that models walk on during a fashion show（时装表演时供模特儿用的）狭长表演台，T 形台

veer towards to change direction towards 转向

vary from sth to sth	to change or be different according to the situation（根据情况）变化，变更，改变
a variety of	several different sorts of the same thing（同一事物的）不同种类，多种式样
intent on/upon sth	giving all your attention to sth 专心；专注
show off	to show people sb/sth that you are proud of 炫耀；卖弄；显示
in addition (to sb/sth)	used when you want to mention another person or thing after sth else 除……以外（还）
as well (as sb/sth)	in addition to sb/sth; too 除……之外；也；还
make sth (from/out of sth)	to create or prepare sth by combining materials or putting parts together 制造；做；组装
replace sb/sth (with/by sb/sth)	to remove sb/sth and put another person or thing in their place（用……）替换；（以……）接替

Reading Comprehension

Understanding the text

Answer the following questions.

1. Who led the trend of wearing white wedding dresses according to the passage?
2. Before Queen Victoria, under what circumstances did people often wear white dresses?
3. What did Queen Victoria's wedding dress look like?
4. Why did she choose the colour white for her wedding dress?
5. Why did she eschew the heirloom jewels, heavy fabrics and rich colours?
6. According to the passage, what do white wedding dresses symbolise?
7. How did Hollywood and celebrity weddings help cement the notion that marriage demanded a white dress?
8. Is white the only choice for wedding dresses? Why or why not?

Critical thinking

Work in pairs and discuss the following questions.

1. Are you willing to choose a colour other than white for your wedding dress? Why or why not?
2. Based on Paragraphs 8–11, how do you understand the relationship between wedding dresses and fashion?

Research project

According to Text A, you may realise that wedding dresses have evolved and changed in terms of colours and materials as time goes by. Choose one wedding dress and introduce its features and stories in class.

Language Enhancement

Words in use

Fill in the blanks with the words given below. Change the form when necessary. Each word can be used only once.

distinctive	symbolism	initiate	aftermath	prosperity
signify	celebrity	cement	intricate	embellishment

1. The watch mechanism is extremely ________ and very difficult to repair.
2. The architect was asked to add some sculptural ________ to thc building design.
3. The ________ of every gesture will be of vital importance during the short state visit.
4. The number 30 on a road sign ________ that the speed limit is 30 miles an hour.
5. She vividly recalled the terrible hunger and poverty she experienced during the final years of the war and its ________ .
6. A ________ feature of qualitative methods is the flexibility of research designs, particularly where ethnographic methods using a range of techniques are

involved.

7. A country's future ________ depends, to an extent, upon the quality of education of its people.
8. He signed his first contract with Universal, changed his name and became a ________ almost overnight.
9. The university's exchange scheme has ________ its links with many other academic institutions.
10. The automaker ________ a programme to improve the recyclability of its automobiles at the end of their useful life.

Banked cloze

Fill in the blanks by selecting suitable words from the word bank. You may not use any of the words more than once.

A. bridal	F. instituted	K. effect
B. reign	G. edicts	L. relaxed
C. previous	H. based	M. intricate
D. patterns	I. initiated	N. monarch
E. signifying	J. colours	O. enforcing

China may be the first place where brides were expected to wear a particular colour. During the 1. ________ of the Zhou Dynasty some three thousand years ago, brides and their bridegrooms both donned sober black robes with red trim, worn over a visible white undergarment. The wearing of specific 2. ________ and designs was not reserved for weddings. Zhou rulers 3. ________ strict clothing laws that dictated what could be worn, by whom, and when, 4. ________ on profession, social caste, gender, and occasion. These rules were still in 5. ________ by the start of the Han Dynasty, around 200 BCE, when brides and bridegrooms still both wore black. The Hans were purportedly less strict in 6. ________ clothing edicts, but nevertheless prescribed that certain colours be worn at certain times of the year: green in spring, red in summer, yellow in autumn, and black in winter.

By the seventh century, during the reign of the Tang Dynasty, with clothing 7. ________ further loosened, it became fashionable for brides to wear green to their weddings—perhaps as a nod to the springtime clothing of the 8. ________ Han period—while their bridegrooms typically wore red. A more 9. ________ social order led to more diverse and experimental fashions, with women wearing short dresses and even traditional menswear in their daily lives. The Tang Dynasty ruled during a period of much immigration and cultural influence that flowed from China to both Japan and the Korean peninsula, and the fashion influences from the Tang period can still be seen in some traditional Japanese and Korean 10. ________ fashions today, both in colour and in form.

Expressions in use

Fill in the blanks with the expressions given below. Change the form when necessary. Each expression can be used only once.

in addition to	intent on	veer towards	make from
vary from...to...	show off	as well as	replace...with...

1. Against this backdrop, the big thinkers about today's media revolution tend to ________ extremes of optimism or pessimism.
2. Like the monthly payments of a mortgage, monthly car payments are divided between paying principal and interest, and the amounts dedicated to each ________ payment ________ payment.
3. Watches, jewelry, apparel, cars and wine are good ways to ________ and gain respect.
4. ________ his apartment in Manhattan, he has a villa in Italy and a castle in Scotland.
5. She is a talented musician ________ a photographer.
6. Creative thinking should be prized, and so should the gains that can be ________ it.
7. The nutritionist says people should ________ white rice and other refined carbohydrates ________ whole grains whenever possible.

8. I was so ________ my work that I didn't notice the time.

Translation

I. Translate the following paragraph into Chinese.

Plaid is one of the world's widest-spread and most-recognizable patterns, appearing on everything from shirts and scarves, to kilts and hats, to rugs and wall-hangings. Originating in northern Scotland, the iconic fabric has existed for thousands of years, and been worn by millions. Technically, "plaid" isn't the correct word for the pattern that's usually known by that name—the correct word for the ancient patterns would be "tartan", with "plaid" meaning a type of heavy cloth cloak worn to protect travelers from the driving snow and pouring rain of Scottish winter. Individual Scottish clans have their own associated tartan pattern, with the history of each quite literally woven into the fabric. Weavers making wool garments would have had only a limited number of dyes and colours available to them, and out of that palette, would have created some of the only possible patterns, which later became artistically elaborated and passed down from generation to generation. As the patterns became more and more associated with the regions they originated from, the tartans grew to symbolise clans from those regions. In the 18th century, tartans were first used as a military uniform in the Scottish rebellion against England, and after the rebellion was defeated, the patterns were banned for almost a century.

II. Translate the following paragraph into English.

结婚礼服是结婚仪式及婚宴时新娘穿的服饰，其颜色和款式体现了特定文化背景和时装潮流。随着中西文化的不断交融，时下很多年轻人也会举行西式婚礼。新娘在婚礼上穿着白色婚纱，因为白色被认为是纯洁的象征。然而，在中式婚礼中，新娘通常穿着红色礼服或旗袍，多绣有代表美好寓意的图案，如凤凰、牡丹、祥云等，而红色象征着好运、长寿和幸福，蕴含了对新人的美好祝福。

Paragraph Writing

How to Develop a Paragraph—Comparison/Contrast

Comparison/Contrast is a method of developing a paragraph or an essay. It helps readers reach a critical decision. It could be a comparison and contrast of two products, two objects, two things, or two issues.

Comparison stresses the likenesses and similarities between two people, places, things, or abstractions. Comparison is used to explain two items by giving information about both and comparing their characteristics, or to clarify something unfamiliar by comparing it to something familiar. Figures of speech such as simile, metaphor and analogy can be used in a comparison paragraph or essay to show similarities in things that are basically different. There are a series of transitions that can be used in a comparison paragraph: *similarly, as, like, likewise, both, as ... as, have ... in common,* etc.

Contrast deals with differences to show how different two people, places, things, or abstractions are. Contrast is used to help readers better understand both of the two objects and to further make judgments about the two objects. There are a series of transitions that can be used in a contrast paragraph: *unlike, while, whereas, however, nevertheless, but, in contrast to, on the contrary*, etc.

Both **point-by-point method** and **subject-by-subject method** can be used in a comparison/contrast paragraph or essay. In **point-by-point** comparison/contrast, writers deal with a series of features of two subjects, and then present their comparison/contrast, discussing all points successively. In **subject-by-subject** comparison/contrast, writers first discuss one subject thoroughly, and then moves on to another.

The following are the formats of developing a paragraph by point-by-point method and subject-by-subject method:

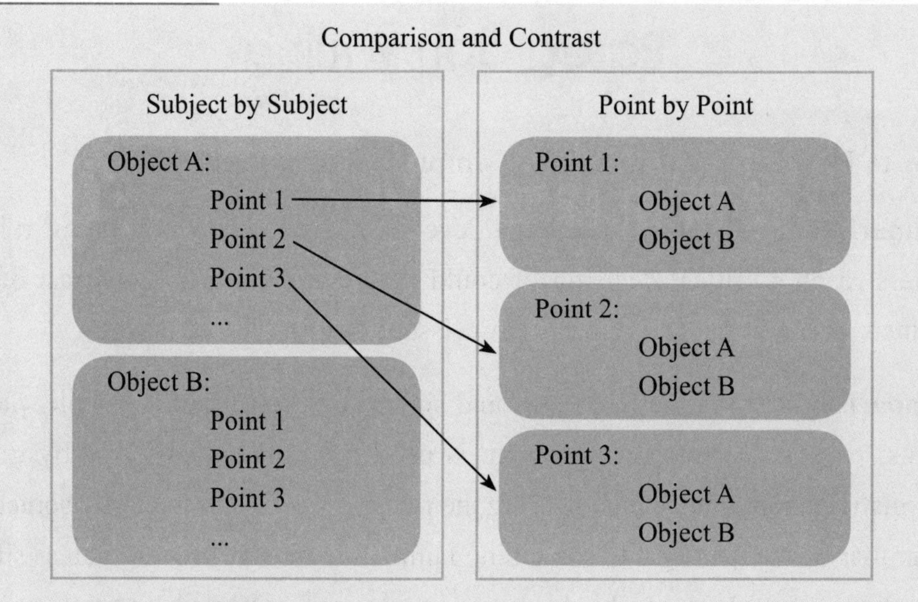

Here are two paragraphs illustrating comparison and contrast respectively. The writer discusses the similarities and differences between the present and previous means of communication. Can you recognise how many transitions are used in each paragraph? And how does the writer organise each paragraph?

Example 1: A comparison paragraph

*Before the advent of computers and modern technology, people communicating over long distances used traditional means such as letters and the telephone. The present and previous means of communication are similar in their general form. One similarity between current and previous methods of communication relates to **the form of communication**. In the past, **both** written forms such as letters were frequently used, in addition to oral forms such as telephone calls. **Similarly**, people nowadays use both of these forms. Just as in the past, written forms of communication are prevalent, for example via email and text messaging. **In addition**, oral forms are still used, including the telephone, mobile phone, and voice messages via instant messaging services.*

Example 2: A contrast paragraph

Before the advent of computers and modern technology, people communicating over long distances used traditional means such as letters and the telephone. The present and previous means of communication differ in their speed and the range of tools available. The most notable of which is speed. This is most evident in relation to written forms of communication. In the past, letters would take days to arrive at their destination. ***In contrast****, an email arrives almost instantaneously and can be read seconds after it was sent. Another significant difference is the range of communication methods. Fifty years ago, the tools available for communicating over long distances were primarily the telephone and the letter.* ***By comparison****, there are a vast array of communication methods available today. These include not only the telephone, letter, email and text messages already mentioned, but also video conferences or mobile phone apps, and social media.*

Choose one of the following topics and write a comparison/contrast paragraph of 80–100 words.

1. High school life and college life
2. Online shopping and traditional shopping
3. Watching a film and reading a novel

Text B

Story of the Suit: The Evolution of Menswear

1 The term "suit" is derived from the French term "suivre" which means "to follow". In other words, the jacket follows the pants or vice versa. So, a suit is a **combination** of a jacket and a pair of pants in a matching fabric. It's not just the colour of the **garments** that is the same, but also the fabric **composition**. It is **undisputed** that the suit embodies **seriousness**, respect and awareness for style. A men's suit is also a sign of **adaptability** and **conservatism**. But where does the suit actually come from?

2 It is the **etiquette** that has been **steadily maintained** since the 17th century, making the suit the most timeless garment ever since. Even during

the last hundreds of years, the suit has only changed in terms of details, but never in the way that is **valid** for current trends in fashion. Perhaps it is also because one can no longer recognise the cultural, **ethnic** or **religious** differences of a man, if he wears a suit. But how can it be that this kind of clothing **persistently** survived and will most certainly remain one of the most important garments in a men's wardrobe in the future?

3 The 17th century was characterised by all kinds of **flamboyance** and **exaggerations**, even in terms of fashion. One of the main reasons was that the **nobility** wanted to properly present its wealth and all luxury it possessed. Think of all the elaborate **corsets**, the intricate embroidery, the **infinite splendour** of colour and the large, **sumptuous** jewelry. Showing off was what it was all about.

4 However, this was different for Charles II, the King of England, who put an end to this. Inspired by King Louis XIV from Versailles in France, he **declared** in 1666 that the royal court would ban **showy** jewelry and the typical **ruff**. Instead, it should be about **elegant** simplicity, about beautiful fabrics with matching cuts. So first the long **vest** was launched, as well as the scarf and a simple shirt with breeches which was paired with a knee length **overcoat**.

5 Of course, this was not the end of the **wigs** or even the bright colours. This only happened during the revolution in the 18th century where the eye-catching dresses of noble people were hated and **despised** more and more. After all, "normal" citizens were not able to wear anything like this themselves and started spreading their own fashion. This consisted of neat and simple colours as well as simple fabrics and comfortable clothes.

6 In England, one could observe these changes even more clearly, but there it was initially based on the **equestrian** fashion from the English upper class. In these circles one would wear long pants with a short vest and a matching white shirt. We also know the overcoats with the typical V-neck, which were especially designed for sitting on a horseback. To make

this horseback riding attire **respectable**, the sporty outfit had to be slightly changed. After all, it was Beau Brummell who took on this task.

7 At first, it was the French who adapted to the new fashion. Later, it reached all of Europe, whereby the simplicity was finally able to **prevail**. **Stylish** people were no longer interested in colourful and **eccentric** clothes. The dark suit was born and was in **vogue** like no other garment. It embodied style and fashion awareness at that time already.

8 By the middle of the 19th century, it was the long **frock coat** and **cutaway**, a formal form of frock coat which were famous. Today, this is commonly known as a **tuxedo**, which was worn by well-dressed men for smoking and drinking with friends. Instead of tuxedo we can also use the term "black-tie." Wearing this garment had a particular reason back then as the tuxedo prevented that the remaining clothes were "smoked in." During the day men would wear a dinner jacket whereas the tuxedo was reserved for the evening, as is still the case today.

9 However, it came to a **turnaround** because men's fashion was not supposed to put itself in the front of women's fashion at that time. Probably this was due to economic reasons. Therefore, the high-quality suit was placed rather in the background. This led to its mass production and it becoming cheaper and cheaper. Due to this development, these neat garments became also **accessible** for people from the lower class. Hence, the suit could establish itself quickly as the formal business attire for all classes.

10 Men did not simply wear anything. They rather adhered to the etiquette and the standards which **established** themselves in this elegant world. For formal occasions, men would wear a cutaway, which was a garment for men developed from the frock coat. If a reception took place in the evening, the perfect choice was a tailcoat. This is a garment with a waist-length jacket with two knee-length tails at the back. We can also use the term "white-tie" if we refer to a tailcoat. When it comes to daily office

work, then a short frock coat was the standard. Gustav Stresemann developed a special design in 1925 which finally turned the suit into an all-rounder. After all, as Foreign Minister, he did not always have the time to change his clothes constantly. Therefore, he simply combined a black jacket with gray pants and created an everyday look.

11 From then on, well-known **politicians** wore the Stresemann look and **thereby** inspired the Americans. Combining a dark jacket with a pair of lighter pants created a new trend. However, the cut of the suit and the actual shape remained unchanged which is the case until today. The same applies to wearing all those well-known men's accessories such as the tie. The tie, also referred to as "**cravat**," is not an **accessory** which was invented to combine it with a suit in the first place. Actually, its first form goes back to the year 200 CE. Even then, people were already wearing linen scarves around their necks. The **predecessor** of today's tie was worn by Croatian soldiers in the 17th century during the Thirty Years' War and helped to distinguish friends from **enemies**.

Notes

Beau Brummell George Bryan Brummell, (1778–1840), English dandy, famous for his friendship with George, Prince of Wales (regent from 1811 and afterward King George IV). Brummell was deemed the leader of fashion at the beginning of the 19th century.

Gustav Stresemann Gustav Stresemann, (1878–1929), chancellor and foreign minister of the Weimar Republic, largely responsible for restoring Germany's international status after World War I. With French foreign minister Aristide Briand, he was awarded the Nobel Prize for Peace in 1926 for his policy of reconciliation and negotiation.

The Thirty Years' War The Thirty Years' War was a 17th-century religious conflict fought primarily in central Europe. It remains one of the longest and most brutal wars in human history, with more than 8 million casualties resulting from military battles as well as from the famine and disease caused by the conflict.

In the end, the conflict changed the geopolitical face of Europe and the role of religion and nation-states in society.

Charles II, the King of England 英格兰国王查理二世

King Louis XIV 法国国王路易十四

New words and phrases

derive /dɪˈraɪv/ *v.* **derive from sth | be derived from sth** to come or develop from sth 从……衍生出；起源于；来自

combination /ˌkɒmbɪˈneɪʃn/ *n.* [C] two or more things joined or mixed together to form a single unit 结合体；联合体；混合体

garment /ˈɡɑːmənt/ *n.* (*formal*) a piece of clothing （一件）衣服

composition /ˌkɒmpəˈzɪʃn/ *n.* [U] the different parts which sth is made of; the way in which the different parts are organised 成分；构成；组合方式

undisputed /ˌʌndɪˈspjuːtɪd/ *a.* ① that cannot be questioned or proved to be false; that cannot be disputed 不容置疑的；毫无疑问的；不可争辩的 ② that everyone accepts or recognises 广为接受的；公认的

seriousness /ˈsɪəriəsnəs/ *n.* [U, sing.] the state of being serious 严重；认真；严肃

adaptability /əˌdæptəˈbɪləti/ *n.* [U] the quality of being able to change or be changed in order to deal successfully with new situations 适应性；适应能力

conservatism /kənˈsɜːvətɪzəm/ *n.* [U] the tendency to resist great or sudden change 保守；守旧

etiquette /ˈetɪket/ *n.* [U] the formal rules of correct or polite behaviour in society or among members of a particular profession （社会或行业中的）礼节，礼仪，规矩

steadily /ˈstedəli/ *ad.* ① gradually and in an even and regular way 稳步地；持续地；匀速地 ② without changing or being interrupted 稳定地；恒定地

maintain /meɪnˈteɪn/ *v.* to make sth continue at the same level, standard, etc. 维持；保持

valid /ˈvælɪd/ *a.* ① that is legally or officially acceptable（法律上）有效的；（正式）认可的 ② based on what is logical or true 符合逻辑的；合理的；有根据的；确凿的

ethnic /ˈeθnɪk/ *a.* connected with or belonging to a nation, race or people that shares a cultural tradition 民族的；种族的

religious /rɪˈlɪdʒəs/ *a.* connected with religion or with one particular religion 宗教的

persistently /pəˈsɪstəntli/ *ad.* ① in a way that shows that you are determined to do something despite difficulties, especially when other people are against you and think that you are being annoying or unreasonable 执著地；不屈不挠地；坚持不懈地 ② in a way that continues for a long period of time, or that is repeated frequently, especially in a way that is annoying and cannot be stopped 连绵地；持续地；反复出现地

flamboyance /flæmˈbɔɪəns/ *n.* ① the fact of being different, confident and exciting in a way that attracts attention 炫耀；卖弄 ② the fact of being brightly coloured and likely to attract attention 艳丽；绚丽夺目

exaggeration /ɪɡˌzædʒəˈreɪʃn/ | *n.* [C] [usually sing., U] a statement or description that makes sth seem larger, better, worse or more important than it really is; the act of making a statement like this 夸张；夸大；言过其实

nobility /nəʊˈbɪləti/ | *n.* (**the nobility**) [sing.+sing./pl.v.] people of high social position who have titles such as that of duke or duchess 贵族

corset /ˈkɔːsɪt/ | *n.* [C] a piece of women's underwear, fitting the body tightly, worn especially in the past to make the waist look smaller （尤指旧时妇女束腰的）紧身内衣

infinite /ˈɪnfɪnət/ | *a.* ① very great; impossible to measure 极大的；无法衡量的 ② without limits; without end 无限的；无穷尽的

splendour /ˈsplendə(r)/ | *n.* [U] grand and impressive beauty 壮丽；雄伟；豪华；华丽

sumptuous /ˈsʌmptʃuəs/ | *a.* (*formal*) very expensive and looking very impressive 华贵的；豪华的；奢华的

declare /dɪˈkleə(r)/ | *v.* ① to say sth officially or publicly 公布；宣布；宣告 ② to state sth firmly and clearly 表明；宣称；断言

showy /ˈʃəʊi/ | *a.* (*often disapproving*) so brightly coloured, large or exaggerated that it attracts a lot of attention 显眼的；艳丽的；花哨的

ruff /rʌf/ | *n.* [C] a wide stiff white collar with many folds in it, worn especially in the 16th and 17th centuries 飞边（尤指盛行于16和17世纪的白色轮状皱领）

elegant /ˈelɪɡənt/ *a*. (of clothes, places and things 衣服、地方及物品) attractive and designed well 漂亮雅致的；陈设讲究的；精美的

vest /vest/ *n*. [C] a short piece of clothing with buttons down the front but no sleeves, usually worn over a shirt and under a jacket, often forming part of a man's suit（西服的）背心

breeches /ˈbrɪtʃɪz/ *n*. [pl.] short trousers/pants fastened just below the knee（裤脚束于膝下的）半长裤，马裤

overcoat /ˈəʊvəkəʊt/ *n*. [C] a long warm coat worn in cold weather 长大衣

wig /wɪɡ/ *n*. [C] a piece of artificial hair that is worn on the head, for example to hide the fact that a person is bald, to cover sb's own hair, or by a judge and some other lawyers in some courts of law 假发

despise /dɪˈspaɪz/ *v*. to dislike and have no respect for sb/sth 鄙视；蔑视；看不起

equestrian /ɪˈkwestriən/ *a*. connected with riding horses, especially as a sport 马术的

respectable /rɪˈspektəbl/ *a*. ① considered by society to be acceptable, good or correct 体面的；得体的；值得尊敬的 ② fairly good; that there is not reason to be ashamed of 相当好的；不丢面子的

prevail /prɪˈveɪl/ *v*. to exist or be very common at a particular time or in a particular place 普遍存在；盛行；流行

stylish /ˈstaɪlɪʃ/ *a*. (*approving*) fashionable; elegant and attractive 时髦的；新潮的；高雅的；雅致的

eccentric /ɪk'sentrɪk/ *a*. considered by other people to be strange or unusual 古怪的；异乎寻常的

vogue /vəʊg/ *n*. [C, usually sing., U] ~ **(for sth)** a fashion for sth 流行；时髦；风行；风尚

frock /frɒk/ *n*. [C] (*old-fashioned*) (*especially BrE*) a dress 连衣裙；女装

frock coat *n*. [C] a long coat worn in the past by men, now worn only for special ceremonies 男长礼服；佛若克男礼服大衣

cutaway /'kʌtəweɪ/ *n*. [C] a black or grey jacket for men, short at the front and very long at the back, worn as part of morning dress （男子日间穿的黑色或灰色的）晨燕尾服，常礼服

tuxedo /tʌk'si:dəʊ/ *n*. [C] a dinner jacket and trousers/pants, worn with a bow tie at formal occasions in the evening（男式，配蝶形领结的）成套无尾晚礼服

turnaround /'tɜ:nəraʊnd/ *n*. [C] a complete change in sb's opinion, behaviour, etc. （观点、行为等的）彻底转变

accessible /ək'sesəbl/ *a*. that can be reached, entered, used, seen, etc. 可到达的；可接近的；可进入的；可使用的；可见到的

establish /ɪ'stæblɪʃ/ *v*. ① to start or create an organisation, a system, etc. that is meant to last for a long time 建立；创立；设立 ② ~ **sb/sth/yourself (in sth) (as sth)** to hold a position for long enough or succeed in sth well enough to make people accept and respect you 确立；使立足；使稳固 ③ to make people accept a belief, claim, custom, etc. 获得接受；得到认可

politician /ˌpɒləˈtɪʃn/ *n*. [C] ① a person whose job is concerned with politics, especially as an elected member of parliament, etc. 从政者；政治家 ② (*disapproving*) a person who is good at using different situations in an organisation to try to get power or advantage for himself or herself 政客；见风驶舵者；投机钻营者

thereby /ˌðeəˈbaɪ/ *ad*. (*formal*) used to introduce the result of the action or situation mentioned 因此；由此；从而

cravat /krəˈvæt/ *n*. [C] a short wide strip of silk, etc. worn by men around the neck, folded inside the collar of a shirt（男用）阔领带

accessory /əkˈsesəri/ *n*. [C] [usually pl.] ① an extra piece of equipment that is useful but not essential or that can be added to sth else as a decoration 附件；配件；附属物 ② a thing that you can wear or carry that matches your clothes, for example a belt or a bag（衣服的）配饰

predecessor /ˈpriːdɪsesə(r)/ *n*. [C] a thing, such as a machine, that has been followed or replaced by sth else 原先的东西；被替代的事物

enemy /ˈenəmi/ *n*. ① [C] a person who hates sb or who acts or speaks against sb/sth 敌人；仇人；反对者② (**the enemy**) [sing.+sing./pl.v.] a country that you are fighting a war against; the soldiers, etc. of this country 敌国；敌军；敌兵

derive from sth/be derived from sth to come or develop from sth 从……衍生出; 起源于; 来自

in other words	expressed in a different way 换句话说；也就是说；换言之
vice versa	used to say that the opposite of what you have just said is also true 反过来也一样；反之亦然
in terms of	used when you are referring to a particular aspect of sth 谈及；就……而言；在……方面
put an end to sth	to make sth stop happening or existing 结束，终止
consist of sth	to be formed from the things or people mentioned 由……组成（或构成）
base sth on/upon sth	to use an idea, a fact, a situation, etc. as the point from which sth can be developed 以……为基础（或根据）
be supposed to do/be sth	to be expected or required to do/be sth according to a rule, a custom, an arrangement, etc.（按规定、习惯、安排等）应当，应，该，须
due to sth/sb	caused by sb/sth; because of sb/sth 由于；因为
lead to sth	to have sth as a result 导致，造成（后果）
adhere to sth	(*formal*) to behave according to a particular law, rule, set of instructions, etc.; to follow a particular set of beliefs or a fixed way of doing sth 坚持，遵守，遵循（法律、规章、指示、信念等）
take place	to happen, especially after previously being arranged or planned（尤指根据安排或计划）发生，进行
turn sb/sth (from sth) into sb/sth	to make sb/sth become sb/sth 使（从……）变成

apply to sb/sth	to concern or relate to sb/sth 有关；涉及
distinguish A from B	to be a characteristic that makes two people, animals or things different 成为……的特征；使具有……的特色；使有别于

Reading Comprehension

Understanding the text

Choose the best answer to each of the following questions.

1. According to Para.1, what does the suit embody?
   A. It embodies humour, adaptability and conservatism.
   B. It embodies seriousness, respect and awareness for style.
   C. It embodies seriousness, respect and awareness for style, adaptability and conservatism.
   D. It embodies humour, respect and awareness for style, adaptability and open-mindedness.
2. What were the characteristics of fashion in the 17th century?
   A. Flamboyance and exaggerations.
   B. Seriousness and conservatism.
   C. Humour and open-mindedness.
   D. Simplicity and purity.
3. Which of the following was not used by the nobility to show off during the 17th century?
   A. The intricate embroidery.
   B. The infinite splendour of colour.
   C. The large, sumptuous jewelry.
   D. The gorgeous tuxedo.
4. What does the elegant menswear launched by Charles II look like?
   A. A short vest, the scarf, a simple shirt, paired with a knee length overcoat.
   B. A long vest, the scarf, a simple shirt with breeches, paired with a knee length overcoat.

C. A long vest, the scarf, a simple shirt, paired with a knee length overcoat.

D. A long vest, the scarf, a simple shirt with breeches, paired with a mid-calf length overcoat.

5. What is the difference between "black-tie" and "white-tie"?

A. "Black-tie" refers to a tuxedo while "white-tie" refers to a tailcoat.

B. "Black-tie" refers to a tailcoat while "white-tie" refers to a tuxedo.

C. "Black-tie" refers to a tuxedo while "white-tie" refers to a suit.

D. "Black-tie" refers to a suit while "white-tie" refers to a tuxedo.

Critical thinking

Read the text carefully and try to find out the key information to fill in the table of the evolution of menswear.

Time	**People who influenced menswear**	**The evolution of menswear**
in 1666	1.________	the 2.________, the 3.________, a simple 4.________ with 5.________, a 6.______
in the 18th century	7.________	long 8.________ with a short 9.______ and a matching 10.________, the overcoats with the typical 11.________
by the middle of the 19th century		a 12.________: the long 13.________ and 14.________ a 15.________; a garment with a 16.______ with two 17.______ at the back
in 1925	18.________	a 19.__________ combined with 20.______

Language Enhancement

Words in use

Fill in the blanks with the words given below. Change the form when necessary. Each word can be used only once.

combination	derive	adaptability	valid	accessible
composition	infinite	respectable	predecessor	persistently

1. He ________ an enormous amount of satisfaction from helping those in need.
2. New technologies work particularly well when they are used in ________ with traditional classroom learning.
3. The company said it plans to change the ________ of its board of directors, creating a board composed predominantly of independent directors.
4. Overseas experience shows ________ and an ability to cope with foreign languages.
5. For the experiment to be ________ , it is essential to record the data accurately.
6. Officials have expressed concern that ________ high oil prices could trigger higher inflation.
7. With ________ patience, she explained the complex procedure to us.
8. It was hard to find a clean shirt that looked ________ enough to be seen in.
9. Public areas and buildings are now more ________ to people with disabilities.
10. He faced the same kind of problems as his ________ .

Expressions in use

Fill in the blanks with the expressions given below. Change the form when necessary. Each expression can be used only once.

adhere to	consist of	put an end to	derive from
lead to	base on	distinguish from	in terms of

1. The complex style of the medieval wooden church carvings and the skills used to make them almost certainly to ____________ the ancient Viking tradition.
2. For the adult learner, the technology has the capability of freeing us ____________ time and space.
3. Innovations usually ____________ some form of recombination of other, existing sources of knowledge.
4. However, such cooperation must ____________ trust, negotiation, equality and mutual benefits. It must proceed from actual facts, rather than deductions.
5. Many governments have ____________ some of the very necessary regulations on cross-border flows of capital, labour and goods, with poor results.

6. Reducing speed limits should ____________ fewer deaths on the roads.

7. The bank ____________ all rules regarding the opening and closing of accounts.

8. They look so similar that it's often difficult to __________ one __________ the other.

Sentence structure

***I*. Complete the following sentences by translating the Chinese into English, using "one of the (main) reasons is/was that ..." structure.**

Model: The 17th century was characterised by all kinds of flamboyance and exaggerations, even in terms of fashion.__________________ ____. (主要原因之一是贵族想要适当地展示其财富和所拥有的奢侈品。)

→ The 17th century was characterised by all kinds of flamboyance and exaggerations, even in terms of fashion. One of the main reasons was that the nobility wanted to properly present its wealth and all luxury it possessed.

1. Travelling overseas is becoming more and more accessible in our daily life. ____ ____ (主要原因之一是英语的广泛使用。)

2. Traffic congestion is a common phenomenon in cities. ______________________ ____ (主要原因之一是车辆数量增长的速度远快于道路建设的速度。)

3. Some students rarely ask questions in math class.___________________________ ____ (其中一个原因是他们害怕丢脸或令人失望。)

***II*. Rewrite the following sentences by using "It is undisputed that...".**

Model: Indisputably, the suit embodies seriousness, respect and awareness for style.

→ It is undisputed that the suit embodies seriousness, respect and awareness for style.

1. Indisputably, Steve Jobs has long-term vision to introduce disruptive (创新的) products to the market place.

____

____.

2. Indisputably, they are the most famous filmmakers in China.

____

____.

3. Indisputably, the application of open spandrel (拱肩) design makes the bridge more exquisite.

____

____.

Cowboy Hats in American History

1 A cowboy hat is defined as an extensive overflowed hat with a huge lenient crown. It is sometimes referred to as ten-gallon hat. It is well designed to have a protruding crown and a wide brim, a shape that can be modified by the one wearing it to suit specific conditions such as fashion or protection against weather. It is a typical hat in the western countries with Mexicans being the most famous people recognised for wearing it. The cowboy hat was common among the people working in ranches in the western and southern United States, country music singers and ranchero musicians in Mexico as well as the participants of the rodeo circuit from North America.

2 One cannot really define a cowboy without mentioning their hats. The hat is probably the utmost defining feature of the cowboy's iconic appearance. There are few items in the history of American culture that carry the same

iconic weight as the cowboy hat. It is the one item of apparel that can be worn in any corner of the world and receive immediate recognition. As the old cowboy saying goes, "It's the last thing you take off and the first thing that is noticed."

3 The history of the cowboy hat is not that old. Before the invention of the cowboy hat, which means before John B. Stetson came along, the cowpunchers of the plains wore castoffs of previous lives and vocations. Everything from formal top hats and derbies to leftover remnants of the civil War headgear, to tams and sailor hats, were worn by men moving westward.

4 The cowboy hat is the most recognised attire in the western countries, but it did not initially have the round, curved brim as those of today. The first cowboy hat was designed by John B. Stetson back in 1865, during the time of American civil war. John Stetson was a famous hat manufacturer and he had roots from Philadelphia. He designed the first cowboy hat, and it was referred to as the "Boss of the Plains" by that time. The hat was made using fine fur from beaver, rabbit and other small animals with fine hair. The hat was perfect for the demands of the western countries, making it gain popularity among the people.

5 The "Boss of the Plains" has undergone different modifications over the years to become the cowboy hat we have now. The Mexicans changed the current cowboy hat in the 19th century. The Mexicans redesigned the hat and made it have a tall crown to provide insulation and made the brim wider so as provide shade from the hot and sunny climates of Mexico. They curved the edges upwards so as not to interfere with the rope. The western people did not have a standard headwear, and with the advancement of the cowboy hat, the western nations adapted to wearing it to favour their different needs. Consequently, the hat gained so much popularity prompting Lucius Beebe to name it as the hat that won the west.

6 Today's cowboy hat has remained basically unchanged in construction and design since the first one was created in 1865. As the story goes, John B. Stetson and some companions went west to seek the benefits of a drier climate. During a hunting trip, Stetson amused his friends by showing them how he could make cloth out of fur without weaving.

7 After creating his "fur blanket," Stetson fashioned an enormous hat with a huge brim as a joke, but the hat was noted to be big enough to protect a man from sun, rain, and all the rigours the outdoors could throw at him. Stetson decided to wear the hat on his hunting trip, and it worked so well that he continued wearing it on his travels throughout the West. In 1865, he began to produce the first incarnation of his big hats in number, and before long, Stetson was considered the maker of this newfangled headwear, the cowboy hat. The original Stetson hat sold for five dollars.

8 Shortly after the turn of the century, the cowboy hat, although still in its infancy, nevertheless infused its wearer with a singular link to the history of the wild and woolly West. Even after the wild aspect of the West was somewhat tamed, the cowboy hat never really lost its ability to lend that reckless and rugged aura to its wearer.

9 The cowboy hat has since then spread to different parts of the planet and is worn for different reasons. Some wear it for weather conditions while others wear it due to its popularity. It has gained so much popularity and they are popular among presidents, musicians, and other notable figures.

10 Nowadays, some cowboy hats are made of straw, but most are still made with felt. Felt is an unusual material that is made of thousands of short animal-fur fibres and therefore tends to twist together when kneaded in hot water and steam. It is also excellent for repelling water and cold temperatures. Cowboy hats can be made of fur felt or wool felt. Fur felt cowboy hats are attractive and lightweight; they can also withstand wet weather conditions without losing their shape.

11 The best quality and most durable cowboy hats are usually made with beaver fur or rabbit fur. Beaver fur is used for most fur felt hats because it is denser, more dimensionally stable, and more water-repellent than rabbit fur.

12 They are enhanced on the inside using a simple band to make it more stable on the head while some cowboy hats may have strings at the base of the crown. The hats come in different colours with brown, black and beige being the most common colours.

Unit 6

Fashion and Trend

Fashion is not something that exists in dresses only. Fashion is in the sky, in the street, fashion has to do with ideas, the way we live, what is happening.

— *Coco Chanel*

Don't be into trends. Don't make fashion own you, but you decide what you are, what you want to express by the way you dress and the way you live.

— *Gianni Versace*

Pre-Reading Activities

1. Watch the video and choose the best answer to each of the following questions.

(1) What was baffling to the female boss?

A. The employees were not fully prepared.

B. The advertisers were not present.

C. The dresses were not beautiful.

(2) What did the man mean by saying "I am on board"?

A. He meant that he totally agreed.

B. He meant that he was going to take a flight.

C. He meant that he was on a ship.

(3) Why did the young lady laugh about the belts?

A. Because she thought those people were funny.

B. Because she couldn't tell the differences between the two belts.

C. Because they were telling a joke.

(4) Why was the boss angry with the word "stuff"?

A. Because she didn't think it was appropriate to describe the fashion industry.

B. Because she thought it was a bad word.

C. Because she didn't think the girl was respectful of the fashion industry.

(5) What did the boss mean by saying "You're wearing a sweater that was selected for you by the people in this room" ?

A. She meant that they selected the sweater.

B. She meant that the sweater perfectly suited the girl.

C. She meant that fashion was influencing everyone.

2. Watch the video again and discuss the following questions.

(1) How many colours did the boss mention to describe the blue sweater? What are they?

(2) Why did the boss mention the colours and the designers?

(3) Do you agree that fashion has nothing to do with the girl? Why or why not?

Text A

Fashion: What does It Mean?

1 What is fashion? As someone with a keen eye for fashion, a **passion** for style and aims of working at a fashion magazine, this is something that has been **running through** my mind since I fell in love with all things of fashion **journalism**.

2 Now, I've always loved fashion, ever since I was little, even more now as a 20-year-old. Over the years it has evolved into so much more, with every **issue** of *Vogue* I read, every fashion article I write and fashion event I report on. I truly see and understand the importance of fashion in this 21st century world.

3 Of course, it's not that much of a simple question to answer. What fashion might mean to one person can be different to the next. Fashion is an art form, and it's also a religion to some.

4 It's a profession, a **glimpse** into someone's **personality**. It's **playfulness**, a **conception** of **escapism** or an act of **disguise**. But **ultimately**, to me, fashion is an individual's statement of self-expression.

5 With **conveying** self-expression through fashion, comes the **articulation** of self-**identity**, showing the world who you are through your fashion choices and using your clothes to tell the world something about yourself. It is for me at least, yet for others clothing is about warmth and comfort or **utility**.

6 Even those who have little or none interest in fashion (which in itself is making a statement), what they choose to wear, whether that be **unconsciously** or otherwise, provides us with a **preview** into that person's personality and **frame** of mind.

7 It is also a means of art that has naturally extended in the innovative world of fashion, which creative artists and designers can use as a platform of self-expression—bringing beauty into the world that like-minded imaginative people can **appreciate**.

8 For designers, simply using fabrics, materials, **architectural** shapes, colours, silhouettes, cuts, fine **detailing** and embellishments are **vital** elements in fabricating the **fascinating**, the **pioneering** and the avant-garde.

9 Fashion allows designers to impact the world we live in by physically creating fashionable **manifestations** and statements of expression, of which were once mere mental concepts.

10 **Fundamentally**, fashion is a **discourse**, an unspoken language of visual signs and symbols, yet is still a form of communication, **via** the eyes rather than **verbally**.

11 It is much like viewing a painting. Fashion is a form of artistic expression in which you can see a designer's **vision** of what she or he is **endeavouring** to express through their collections and you can see an individual's passion for style when he or she walks down the street and gets people's heads turning.

12 Without fashion, there wouldn't be the inspiring **fashionistas** we all know and love, nor would they have contributed to the industry with their **pivotal** pieces and change the history of fashion forever.

13 Without fashion there would be no Giorgio Armani with his clean tailored menswear, no Vivienne Westwood to bring modern **punk** and new wave fashions into the mainstream; there would be no Anna Wintour to push the boundaries of fashion journalism; no Coco Chanel to create the most **influential** garment of all time, the little black dress; there would be no Christian Louboutin high heels with the **gorgeous signature** shiny, red-**lacquered** sole, no Marc Jacob watches and handbags to **drool** over, no

Andre Leon Talley to grace the FROWs of many London, Milan, New York and Paris fashion shows, no Burberry trench coat... you get the picture.

14 More specifically, there wouldn't be those important weeks in a year where designers, **glitterati**, bloggers and **aspiring** fashion professionals alike to come together and celebrate fashion in all of its **magnificence**.

15 **Showcasing** over 250 designers to global audiences of influential media and retailers every year, dictating clothing and beauty trends, known for **inspirational** sportswear (New York), **edgy** and avant-garde (London), **exaggerated** yet stylish (Milan) and haute couture (Paris) designs, Fashion Week is one of the most influential and ground-breaking times for all lovers of fashion.

16 Fashion has the power to make us feel something. It acts as an outlet for hope and inspiration for conveying the conceptions we have about ourselves and the image we want to **portray**, or have.

17 **Undeniably**, the fashion industry is a world of imagination, innovation and to some extent, **fantasy**, one that has the ability to **manipulate** our **perception** of the world and, **essentially**, how the world views us.

18 Through the use of the **symbolic** manner of fashion, as individuals, we are proudly stating "this is who I am, this is my culture." Principally, the style world enables an individual to break through the norms and **conventions**, as well, connect with outlet that **reflects** them. Fashion, unquestionably, is an expression of identity.

19 Overall, I have met some of the most amazing people and have made some close friends just through the shared interest of fashion, undoubtedly, it brings people together. I am thankful for the experiences my love for fashion has brought me and I look forward for what my passion might bring me in the future.

Notes

Andre Leon Talley Andre Leon Talley is an American a fashion journalist and creative director and American editor-at-large of *Vogue* magazine. He was the magazine's fashion news director from 1983 to 1987, and then its creative director from 1988 to 1995. He has authored three books including two memoirs and co-authored a book with Richard Bernstein. Talley has also served as international editor of Russian fashion magazine Numéro.

FROW the row of seats closest to the catwalk at a fashion show, considered to be the most prestigious and desirable place to sit; the front row 秀场前排座位

New words and phrases

passion /ˈpæʃn/ *n.* ① [C, U] a very strong feeling of love, hatred, anger, enthusiasm, etc. 强烈情感；激情 ② [C] ~ **(for sth)** a very strong feeling of liking sth; a hobby, an activity, etc. that you like very much 酷爱；热衷的爱好（或活动等）

journalism /ˈdʒɜːnəlɪzəm/ *n.* the work of collecting and writing news stories for newspapers, magazines, radio or television 新闻工作

issue /ˈɪʃuː; ˈɪsjuː/ *n.* ① [C] one of a regular series of magazines or newspapers 一期；期号② [C] an important topic that people are discussing or arguing about 重要议题；争论的问题

glimpse /glɪmps/ *n.* ① ~ **(at sb/sth)**~ **(of sb/sth)** a look at sb/sth for a very short time, when you do not see the person or thing completely 一瞥；一看 ② ~ **(into sth)**~ **(of sth)** a short experience of sth that helps you to understand it 短暂的感受（或体验、体会）

personality /ˌpɜːsəˈnæləti/ *n.* ① the combination of characteristics or qualities that form an individuals distinctive character 性格；个性；人格 ② [U] the qualities of a person's character that make them interesting and attractive 魅力；气质；气度

playfulness /ˈpleɪflnəs/ *n.* ① the quality of being light-hearted or full of fun 趣味性 ② a disposition to find (or make) causes for amusement 兴致

conception /kənˈsepʃn/ *n.* ① [U] the process of forming an idea or a plan 构思；构想；设想 ② [C, U] ~ **(of sth)** ~ **(that** ...) an understanding or a belief of what sth is or what sth should be 理解（认为某事怎样或应该怎样）

escapism /ɪˈskeɪpɪzəm/ *n.* [U] an inclination to retreat from unpleasant realities through diversion or fantasy 逃避现实；解脱方法

disguise /dɪsˈgaɪz/ *v.* ① to change someone's appearance so that people cannot recognize them 假扮；装扮；伪装 ② to hide a fact or feeling so that people will not notice it 掩蔽；掩饰

ultimately /ˈʌltɪmətli/ *ad.* ① as the end result of a succession or process; finally 最终；最后；终归 ② at the most basic and important level 最基本地；根本上

convey /kənˈveɪ/ *v.* ① make an idea impression or feeling known or understandable to someone 表达，传递（思想、感情等）② ~ **sb/sth (from** ...) **(to** ...) to take, carry or transport sb/sth from one place to another 传送；运送；输送

articulation /ɑːˌtɪkjuˈleɪʃn/ *n.* [U] ① (*formal*) the expression of an idea or a feeling in words（思想感情的）表达 ② (*formal*) the act of making sounds in speech or music 说话；吐字；发音

identity /aɪˈdentəti/ *n.* ① the fact of being who or what a person or thing is 身份；本体 ② the characteristics, feelings or beliefs that distinguish people from others 特征；特有的感觉（或信仰）③ [U] ~ **(with sb/sth)** ~ **(between A and B)** the state or feeling of being very similar to and able to understand sb/sth 同一性；相同；一致

utility /juːˈtɪləti/ *n.* [mass noun] the state of being useful, profitable or beneficial 功用；效用；功利

unconsciously /ʌnˈkɒnʃəsli/ *ad.* without realizing or being aware of one's actions; not knowing or perceiving 无意地；不知不觉地

preview /ˈpriːvjuː/ *n.* ① an inspection or viewing of something before it is but or becomes generally known and available 预审；预观 ② a showing of a film, play, exhibition, etc., before its official opening （电影）预告；试映；（展览会）预展

frame /freɪm/ *n.* ① [sing.] (*literary*) the structure constitution or nature of someone or something (人或事物的) 结构，构成，构造；本性 ② a rigid structure that surrounds or encloses a picture, door, windowpan, or similar 画框；门框；边框；框架

appreciate /əˈpriːʃieɪt/ *v.* ① ~ **sb/sth** to recognize the good qualities of sb/sth 欣赏；赏识；重视 ② to be grateful for sth that sb has done; to welcome sth 感激；感谢；欢迎 ③ understand a situation fully; recognize the full implications of 充分理解；充分意识到

architectural /ˌɑːkɪˈtektʃərəl/ *a.* of or relating to the design and construction of architecture 建筑学的；建筑方面的

detailing /ˈdiːteɪlɪŋ/ *n.* [mass noun] small decorative features on a building garment or work of art (建筑物，服装，艺术品的)装饰性细节特征

vital /ˈvaɪtl/ *a.* ① necessary or essential in order for sth to succeed or exist 必不可少的；对……极重要的 ② full of energy and enthusiasm 生机勃勃的；充满生机的；热情洋溢的

fascinating /ˈfæsɪneɪtɪŋ/ *a.* extremely interesting and attractive 极有吸引力的；迷人的；引人入胜的

pioneer /ˌpaɪəˈnɪə(r)/ *v.* develop or be the first to use or a new method of knowledge or activity 开发；开拓；开辟；开创

manifestation /ˌmænɪfeˈsteɪʃn/ *n.* [C, U] ~ (**of sth**) an event action or object that clearly shows or embodies something, especially a theory or an abstract idea 显示；表明；证明（尤指理论或抽象概念）

fundamentally /ˌfʌndəˈmentəli/ *ad.* in every way that is important; completely 根本上；完全地

discourse /ˈdɪskɔːs/ *n.* ① [U] a text or conversation 语篇；谈话 ② a long and serious treatment or discussion of a subject in speech or writing 论文；演讲

via /ˈvaɪə; ˈviːə/ *prep.* ① by way of ; through 经过；通过 ② by means of a particular person, system, etc. 通过；凭借（某人、系统等）

verbally /ˈvɜːbəli/ *ad.* in spoken words and not in writing or actions 口头上（而非书面或行动上）

endeavour /ɪnˈdevə(r)/ *v.* try hard to do or achieving sth 努力；尽力

fashionista /ˌfæʃnˈiːstə/ *n.* ① (*informal*) a designer of haute couture (非正式) 高级女装设计师 ② a devoted follower of fashion 赶时髦的人；超级时尚迷

pivotal /ˈpɪvətl/ *a.* of crucial importance in relation to the development or success of something else (对事物的发展或成功) 起关键作用的，起中心作用的

punk /pʌŋk/ *n.* ① [mass noun] (also punk rock) a loud, fast-moving and aggressive form of rock music, popular in the late 1970s 朋克摇滚乐 ② (also punk rocker) someone who likes punk music and wears things that are typical of it, such as torn clothes, metal chains, and coloured hair 朋客青年

influential /ˌɪnfluˈenʃl/ *a.* having great influence on someone or something 有影响的；有实力的

gorgeous /ˈgɔːdʒəs/ *a.* ① (of colours, clothes, etc.) with very deep colours; impressive (颜色、衣服等) 绚丽的；灿烂的；华丽的 ② (*informal*) very beautiful and attractive; giving pleasure and enjoyment 非常漂亮的；美丽动人的；令人愉快的

signature /ˈsɪgnətʃə(r)/ *n.* a person's name written in a distinctive way as a firm of identification in authorizing a check or document or concluding a letter 署名，签名

lacquer /ˈlækə(r)/ *v.* ① to cover or coat something with lacquer 涂漆，刷漆 ② (*old-fashioned*) to spray lacquer on your hair 用定型液喷头发

drool /druːl/ *v.* ① ~ **(over sb/sth)** to show in a silly or exaggerated way that you want or admire sb/sth very much （对……）垂涎欲滴，过分痴迷 ② to let saliva (= liquid) come out of your mouth 垂涎；淌口水

glitterati /ˌglɪtəˈrɑːti/ *n.* (*informal*) the fashionable set of people engaged in show or some other glamorous activity (娱乐界或其他充满吸引力的活动中的）时髦人士；知名人士

aspiring /əˈspaɪərɪŋ/ *a.* ① wanting to start the career or activity that is mentioned 渴望从事……的；有志成为……的 ② wanting to be successful in life 有抱负的；有志向的

magnificence /mægˈnɪfɪsns/ *n.* the quality of being magnificent 壮丽，宏伟；壮观，辉煌

edgy /ˈedʒi/ *a.* ① (*informal*) at the forefront of a trend; cutting-edge 前卫的，引领潮流的；尖端的，最先进的 ②tense nervous or irritable 紧张不安的; 烦躁的，易怒的

exaggerate /ɪgˈzædʒəreɪt/ *v.* represent sth as being larger, greater, better or worse than it really is 夸大；夸张

portray /pɔːˈtreɪ/ *v.* ① depict sb or sth in a work of art or literature (在艺术或文学作品中）描述，刻画，描写② ~ **sb/sth** (**as sb/sth**) describe sb or sth in a particular way 将……描写成；给人以某种印象；表现

undeniably /ˌʌndɪˈnaɪəbli/ *ad.* used to emphasise that sth cannot be denied or disputed 不可否认地；无可置疑地

fantasy /ˈfæntəsi/ *n.* [C] ① a pleasant situation that you imagine but that is unlikely to happen 幻想；想象 ② [C] an idea with no basis in reality 空想

manipulate /məˈnɪpjuleɪt/ *v.* ① to control or influence sb/sth cleverly, often in a dishonest way so that they do not realize it（暗中）控制，操纵；影响 ② handle or control (a tool, mechanism, etc.), typically in a skillful manner（熟练地）操作，使用（工具，机械等）

perception /pəˈsepʃn/ *n.* ① a way of interpreting, understanding or regarding sth; a mental impression 观念，认知，感知；看法，印象 ② the ability to see here or become aware of sth through the senses 感知能力，认知能力

essentially /ɪˈsenʃəli/ *ad.* used to emphasise the basic, fundamental, or intrinsic nature of a person, thing, or situation 基本上；本质上

symbolic /sɪmˈbɒlɪk/ *a.* ~ **(of sth)** containing symbols, or being used as a symbol 使用象征的；作为象征的；象征性的

convention /kənˈvenʃn/ *n.* the way in which sth is done that most people in a society expect and consider to be polite or the right way to do it 习俗；常规；惯例

reflect /rɪˈflekt/ *v.* ① to show or be a sign of the nature of sth or of sb's attitude or feeling 显示，表明，表达（事物的自然属性或人们的态度、情感等）② ~ **(on/upon sth)** to think carefully and deeply about sth 认真思考；沉思；反思

run through to spread or be present throughout sth; to permeate or pervade sth 贯穿

a glimpse into a short experience of sth that helps you begin to understand it 对……的一瞥，一窥

provide sb with to give sth to sb or make it available to them, because they need it or want it 为某人提供

get people's heads turning to capture people's attention due to being exceptionally interesting, beautiful, unusual, or novel 引人注目；回头率高

push the boundaries of	to make a new discovery, work of art etc. that is very different from what people have known before, and that changes the way they think 开拓，扩展（新领域，新思路）；挑战极限
of all time	that has ever lived or existed; always 历来；一直，始终；有史以来
drool over	envy without restraint; love unquestioningly and uncritically or to excess; venerate as an idol 痴情；垂涎于；眼馋
act as	function as or act like 担任；担当
to some extent	somewhat; partly; in a limited way or to a limited degree 在一定程度上；从某种程度上讲；从某种程度来讲

Reading Comprehension

Understanding the text

Answer the following questions.

1. What does fashion mean to the author ultimately?
2. What can we know about people from their choices of clothing?
3. How does fashion allow designers to impact the world we live in?
4. Why is fashion a form of visual rather than verbal communication?
5. What can we see about designers through their collections?
6. Why Fashion Weeks are important?
7. According to the author, what is fashion industry?
8. Why is fashion an expression of our identity?

Critical thinking

Work in pairs and discuss the following questions.

1. What does fashion mean to you?

2. How do you understand the remark "Fashion allows designers to impact the world we live in"?

Research project

Have you ever thought about the message your clothes send and what does it say about you? Work in groups of four or five and discuss this question. Then, design some questions and interview some students on campus to find out what messages clothes may say about people.

Language Enhancement

Words in use

Fill in the blanks with the words given below. Change the form when necessary. Each word can be used only once.

disguise	identity	endeavour	influential	convey
aspiring	exaggerate	manipulate	symbolic	conventionally

1. Parents always ________ to set their kids on a path that is the least bumpy, and likely to avoid dangerous confrontation.
2. Avoiding ________ theft can be as simple as reviewing your bank statements promptly and never giving a stranger your PIN number or social security number.
3. A double heart diamond ring is a lovely ________ piece of jewelry that can be used for romantic occasion.
4. But the original point of couture, to have a garment made to ________ your flaws and make your life easier, seems to be getting faded.
5. Just like good newspaper headlines lead readers into the story, the title of each slide should ________ the message of that slide.
6. Although he came from humble beginnings, Abraham Lincoln became one of the most ________ presidents of the United States.
7. It is scarcely possible to ________ the difficulties with which he found himself confronted, but he proved himself more than equal to the task.

8. In this uplifting, yet tragic story about ________ artists, the music stirs your soul and you will leave the theater singing.
9. Learning language is ________ easier for children than for adults, especially when the learner is immersed in the new language.
10. In addition to preset modes, most quality digital cameras allow users to ________ the settings manu.

Banked cloze

Fill in the blanks by selecting suitable words from the word bank. You may not use any of the words more than once.

A. determining	F. exposed	K. editorial
B. define	G. passion	L. including
C. conventional	H. market	M. inspire
D. commercial	I. following	N. afford
E. decide	J. convenient	O. basic

It's hard to 1. ________ fashion simply because it means something different to everyone. It could be someone's 2. ________ to express their creativity. Or it could be some people's way of putting food on the table for themselves or their families. And it may just be a foreign world all around someone but they have no interest in 3. ________ up with trends, designers, and such. Because fashion is such a broad term, there are several factors that break it down season by season and determine what fashion is at the moment.

The first factor is haute couture fashion houses 4. ________ Chanel, Christian Dior, and Jean Paul Gaultier, just to name a few. Haute couture houses have been leading the way in creating trends and classic, iconic fashion looks.

The media is another 5. ________ factor in establishing fashion trends. Media outlets such as television shows, movies, social media accounts, and magazines create thousands of trends and fads. These forms of media show people millions of different looks and endless possibilities in the way certain items can be worn together. The way an item is placed in a unique fashion 6. ________ may 7. ________ people to try a new look.

Another factor in determining fashion are celebrities. These days we are 8. ________ to celebrities through all kinds of medias, so it's no surprise they play such a big role in fashion. Designers love to use major stars to 9. ________ their brands, so many times they will give current "A-listers" free clothing, shoes, purses, and accessories as a way to promote themselves through these celebrities to the mass public. This makes the public aware of the designers' lines and encourages them to want to purchase the items whether they can 10. ________ to or not.

Expressions in use

Fill in the blanks with the expressions given below. Change the form when necessary. Each expression can be used only once.

little or none	get people's heads turning	act as	a glimpse into
to some extent	push the boundaries of	drool over	of all time

1. Through the words and photos she used to document her life, we get ____________ the tender, subtle, and sometimes poetic heart of hers.
2.The store's tomatoes had ____________ of the flavour I get from eating what I grow in my garden.
3. ____________, the competition of comprehensive national power of a country is the scientific and technological competition.
4. With the right attitude and confidence, you can wear this creative and artful style dress and will ____________.
5. My boyfriend has been ____________ the leading singer during the concert.
6. This research shows that people will continue to ____________ science even through extreme crisis and severe lack of funds.
7. The general manager must ____________ a kind of public relations person as well as the head of the company's management.
8. Beethoven, one of the composers ____________, would not have produced his music without a strong independent self-esteem.

Translation

I. Translate the following paragraph into Chinese.

Haute couture is a world in which the average consumer can only dream of participating. To the French, haute couture is about much more than just beautiful ballgowns; in fact, it's an area protected by French law. What does it take to be involved in such a highly regulated industry? Haute couture means "high sewing" or "high fashion." It's a step above prêt-à-porter（成衣时装）, the designer collections that most fashionistas will be familiar with. In the world of haute couture, workers can spend up to 700 hours creating a single garment, which is designed for an exclusive clientele（客户）of about 2,000 buyers. If something is labeled as haute couture, it means that it is a one-of-a-kind garment that has been custom-created（定制的） for a specific client.

II. Translate the following paragraph into English.

2018年初，代表中国原创力量的国民运动品牌中国李宁登上了纽约时装周。在此次时装秀上，中国李宁以“悟道”为主题，坚持国人“自省、自悟、自创”的精神内涵，从运动视角表达了对中国传统文化和现代潮流时尚的理解。这个创立近30年的老品牌在世界顶级秀场上完美演绎了90年代复古、现代街头文化以及未来运动趋势三大潮流方向，向全世界展现了中国李宁的潮流与创新认知，引发了设计界对于中国元素的关注，让中国潮牌在国际舞台崭露头角。

Pararaph Writing

How to Develop a Paragraph—Cause & Effect

Cause and effect is a method of paragraph development in which a writer analyses the reasons for and/or the consequences of an action, event, or decision. Paragraphs structured as cause and effect explain reasons why something happened or the effects of something. These paragraphs can be ordered as causes and effects or as effects and then causes.

To put it another way, when you give reasons why something happened, you are explaining what causes an effect (Reasons are causes and the thing that happens is the effect.). You will probably come up with one or more causes. Also, when you explain the results of an action, you are explaining the effects of a cause (Results are effects and the thing that occurs is the cause.). Very often, causes have a single effect while multiple effects maybe the result of one single cause.

So, the usual logical pattern we follow when writing such a kind of paragraph is like this:

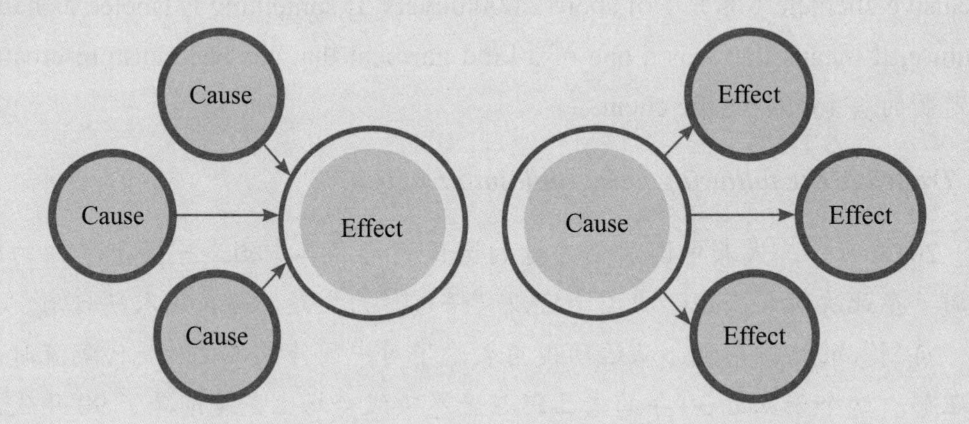

Since a short paragraph is hard to include causes and effects, you are required to put your emphasis on either causes or effects instead of both. In order to develop a good paragraph, you may follow the following steps:

Step 1: Establish your direction. Decide whether you want to talk about causes or effects.

Step 2: Present a clear thesis. Thesis should inform readers of your purpose or intention. Thesis may focus on causes or effects.

Step 3: Follow an organisational pattern. There are two basic ways to organise a cause-and-effect essay: chronological (time) order and emphatic order. Chronological order discusses the causes and effects in the order that they occur. Emphatic order reserves the strongest or most significant cause and/or effect until the end.

Step 4: Use transitions. Transitional words help the reader follow your cause-and-effect analysis.

Step 5: Draw a conclusion. Restate the thesis and reach a conclusion concerning the causes and/or effects.

And then, the right language is important. There are several language formulas that can effectively show cause and effect relationships, so take the time to choose the best ones for your paragraph. As always, vary your sentence structures for a smoother read and use plenty of evidence to write a convincing paragraph, then try some of these phrases to take your cause and effect arguments to the next level.

Cause language

· There are several reasons for ...

· The main factors are ...

· The first cause is ...

· [Cause] leads to or might lead to [effect] ...

· This often results in...

Effect language

· Before [cause] ... Now [effect] ...

· One of the results/outcomes of [cause] is ... Another is ...

· A primary effect of [cause] is ...

· [Effect] often occurs as a consequence of [cause].

Now let's take a look at the following examples:

Example 1: *Many people think that they can get sick by going into cold weather improperly dressed; however, illnesses are not caused by temperature—they are caused by germs. So while shivering outside in the cold probably won't strengthen* *your immune system, you're more likely to contract an illness indoors because you will have a greater exposure to germs.*

In the above example, the paragraph explains how germs cause illnesses. The germs are the cause in the paragraph and the illness is the effect.

Example 2: *Students are not allowed to chew gum in my class. While some students think that I am just being mean, there are many good reasons for this rule. First, some irresponsible students make messes with their gum. They may leave it on the bottoms of desks, drop it on the floor, or put it on other people's property. Another reason why I don't allow students to chew gum is because it is a distraction. When they are allowed to chew gum, students are more worried about having it, popping it, chewing it, and snapping it when they are in listening, writing, reading, and learning. This is why I don't allow students to chew gum in my class.*

For this paragraph, the author explains why students are not allowed to chew gum in his class. There are three causes. And the thesis is restated in the end to echo the effect.

Write a paragraph of no less than 60 words on one of the following topics.

1. Automation can be a good thing.
2. Mountain bikes can ride so smoothly on rough surfaces.
3. Under no circumstances will firecrackers be allowed in school.

Fashion Trends: How do They Start?

1 People usually talk a lot about fashion trends and how they always prefer to keep their looks updated. But have you ever wondered how do these

trends start? You may see models on the runway showing new types of patterns and clothing, or famous celebrities **donning** a particular outfit. And then the next thing you know, you are seeing that **peculiar** print everywhere from different clothing stores to your social media **feed**. A **fanaticism** soon takes off that sees every fashion **enthusiast** rushing to get the piece, causing its demand to **surge dramatically**.

2 This happened with **khaki** pants! It took the fashion industry by storm and people everywhere started to go **fanatical** for them mainly because of two solid reasons. First, it looks extremely **fashionable** and second, the pants are really comfortable.

3 Fashion trends though sometimes seem to happen by accident but, in reality, this is not **necessarily** the case. In today's **digital** world, the Internet and social media have **altered** the fashion trend game. In the less-digital decades of the past, fashion trends used to emerge via fashion houses and magazines. However, things are absolutely different now! Here are some ways that influence and **motivate transitions** in fashion trends.

4 Trend Books & Magazines. It's not a surprise to know that most fashion trends emerge from trend books and magazines. Big fashion companies usually **deploy** teams of **professionals** to travel the world and find ideas for future fashion. They do this by observing and by gaining inspiration from different architectures, fabrics, and cultures. Once the team has gained ideas for fashion trends, they **sort** and incorporate them, **compile** the best ones and publish them in either trend books or magazines, which are then sold to the public, spreading awareness of the **forthcoming** trends for some time before efforts are made to commercialise them.

5 Fashion Shows. When it comes to the best ways leading fashion trends, fashion shows are definitely the first to bear the brunt. Runway shows are one of the best choices to **highlight** the next biggest trends of the **imminent** seasons. For example, high-end Fashion Week events **inspire multitudes** of trends. Therefore, many people wait in **anticipation** to see

what the designers have in store for them for the next seasons. The main reason why **various** trends emerge from runway looks is that the outfits that are displayed are based on the **original** and creative ideas of the designers. These events also show couture pieces that are simply beautiful and designed for everyday wear.

6 **Bloggers & Influencers**. Thanks to the **prevalence** of online blogs and social media platforms, both bloggers and influencers have become a major driving force that contributes to the process of trend creation. Such **individuals** often have an **impressive** following who look up to them for ideas and inspiration, including fashion. They are highly influential in their **niche** areas. This is why many designers and brands reach out to influencers so they can **endorse** and market their products. Fashion bloggers and influencers often **present** their own trends by combining their outfits in interesting ways that get people talking. They offer a fresh **perspective** on the fashion industry and their readers **adore** and respect that. They then share their look on different social media platforms which creates even more **buzz** among their audiences.

7 Street Style. Street style refers to the style that you see every day on the streets. These are the looks or outfits that people normally wear in their everyday life. It also can have its impact on other people. The reason why some people might get inspired by the street style is that it doesn't take much effort to **recreate** the look. Also, people tend to follow these trends because they believe it is something they can pull off themselves. Overall, street style is a newer term in the world of fashion, but it shows how much of an impact that anyone can play in the way that fashion is perceived and trends are set! An example that originated from street style is the Coachella **hippie** look.

8 Fashion Capitals of the World. Fashion trends vary all across the world which is why people are often eager to see what is going on in the fashion capitals of the world, i.e. Paris, New York City, London, and Milan. Whenever people, including designers, fashion lovers, bloggers and

magazine editors-in-chief, look out of their **geographic** area to trend-source, they often reach towards these highly **trendy** cities looking for a **distinctive newness**. They then observe what people are wearing before they try to adopt that trend. Also, this is why designers and fashion lovers travel to these cities to know what's **hip** and in fashion.

9 Celebrities. Perhaps one of the **mightiest** driving forces responsible for creating trends is none other than celebrities. This is because they are already very famous and are followed by millions who like to **imitate** everything that they do. Let's take the example of the American reality TV star and entrepreneur, Kylie Jenner. Every time she shares a picture of herself on the photo and video sharing app, Instagram, her fans quickly go wild and try to dress the same way she dresses. There are also several fast fashion companies such as Fashion Nova that design and model some of its outfits based on Jenner's style and market them as a cheaper alternative to what she is originally wearing.

10 Also, since celebrities usually have a global **fanbase**, they normally have more reach as compared to magazines. The fact that celebrities are highly influential is the very reason why huge companies turn to them to be **spokespeople** for their brands and products. There is no doubt that celebrities are style **icons** as many people turn to them for style and fashion inspiration.

11 These are some of the main sources from where fashion trends originate from. Overall, fashion trends from these sources have the opportunity to make a huge **splash** and impact the fashion industry worldwide. A trend usually starts with these key players before it finds its way to everyday people. Also, some of the fashion trends are not **everlasting**. Fashion trends will suddenly become very popular, disappear over time, and then have the ability to be recycled and serve as inspiration again in the years to come. So, keep experimenting and examine different fashion sources carefully in order to stay up to date with the latest fashion trends!

Notes

Coachella hippie look Coachella is a giant annual music festival in Indio, California. The festival has grown tremendously since its first year in 1999. Coachella is held across two weekends every April, although the dates are not always the same. The festival is usually on Friday, Saturday, and Sunday. Although Coachella's roots are all about music, it features art installations and has also become a major fashion hotspot for bohemian-inspired styles and a Mecca for people to watch and capture street style. After the first year, the hippie look emerged and became very strong, as designer began using it as an inspiration for their upcoming collections and it has rolled into being very dominant and lasting season after season.

Fashion Nova Fashion Nova is a Los Angeles-based clothing retailer that specialises in inexpensive streetwear, especially modern, stylish women's apparel for all body types. They cater to Instagram and are known for their savvy on the platform.

New words and phrases

don /dɒn/ *v.* to put on (an article of clothing) 穿上

peculiar /pɪˈkjuːliə(r)/ *a.* different from the usual or normal 独特的；独有的

feed /fiːd/ *n.* an Internet service in which updates from electronic information sources (such as blogs or social media accounts) are presented in a continuous stream 订阅源

fanaticism /fəˈnætɪsɪzəm/ *n.* fanatic outlook or behavior 狂热

enthusiast /ɪnˈθjuːziæst/ *n.* a person who is highly interested in a particular activity or subject 爱好者；热衷者

surge /sɜːdʒ/ *v.* ① move suddenly and powerfully forward or upward 汹涌；奔腾 ② increase suddenly and powerfully 激增

dramatically /drəˈmætɪkli/ *ad.* remarkably; strikingly 显著地，剧烈地；戏剧性地；引人注目地

khaki /ˈkɑːki/ *n.* ① a dull yellowish-brown colour 卡其色 ② a hard-wearing fabric of this colour, used especially for military uniforms 卡其布

a. ① of the colour khaki 卡其色的 ② made of khaki 卡其布的

fanatical /fəˈnætɪkl/ *a.* ① filled with excessive and single minded zeal 狂热的；痴迷的 ② obsessively concerned with sth 过分在意的；成癖的

fashionable /ˈfæʃnəbl/ *a.* characteristic of, influenced by, or representing a current popular trend or style 时尚的；时髦的；流行的

necessarily /ˌnesəˈserəli/ *ad.* used to say that sth cannot be avoided 必定地；必然地

digital /ˈdɪdʒɪtl/ *a.* involving or relating to the use of computer technology 计算机技术的；数字的；数码的

alter /ˈɔːltə(r)/ *v.* change or cause to change in character or composition typically in the comparatively small but significant way（使）改变性质或成分

motivate /ˈməʊtɪveɪt/ *v.* stimulate one's interest in or enthusiasm for doing sth 激发某人的积极性

transition /trænˈzɪʃn/ *n.* the process or a period of changing from one state or condition to another 过渡；转变；变革

deploy /dɪˈplɔɪ/ *v.* to spread out, utilise or arrange for a deliberate purpose 为某一目的部署；利用

professional /prəˈfeʃənl/ *n.* a person engaged or qualified in a particular activity 专业人员；专家

sort /sɔːt/	*v.* arrange systematically in groups separate according to the type class, etc. 整理；（按类型，等级）分类
compile /kəmˈpaɪl/	*v.* produce (a list, report or book) by assembling information collected from other sources 汇编；编辑；编撰
forthcoming /ˌfɔːθˈkʌmɪŋ/	*a.* planned for all about to happen in the near future（根据计划）即将来临的；即将发生的
highlight /ˈhaɪlaɪt/	*v.* make visually prominent; emphasise 使突出；使显著；强调
imminent /ˈɪmɪnənt/	*a.* about to happen 即将发生的；迫近的
inspire /ɪnˈspaɪə(r)/	*vt.* ① ~ **sb (to sth)** to give sb the desire, confidence or enthusiasm to do sth well 激励；鼓舞 ② [usually passive] to give sb the idea for sth, especially sth artistic or that shows imagination 赋予灵感；引起联想；启发思考
multitude /ˈmʌltɪtjuːd/	*n.* ① ~ **(of sth/sb)** an extremely large number of things or people 众多；大量 ② **the multitude** (also the multitudes) the mass of ordinary people 群众；大批百姓；民众
anticipation /ænˌtɪsɪˈpeɪʃn/	*n.* ① [U] a feeling of excitement about sth (usually sth good) that is going to happen 期盼；期望 ② [U] the fact of seeing that sth might happen in the future and perhaps doing sth about it now 预料；预期；预见；预计
various /ˈveəriəs/	*a.* ① different from one another; of different kinds or slots 不同的；各种各样的② having many different features 具有多种特征的；多姿多彩的

original /əˈrɪdʒənl/ *a.* ① new and interesting in a way that is different from anything that has existed before; able to produce new and interesting ideas 首创的；独创的；有独创性的 ② used or produced at the creation or earliest stage of something 起初的；原来的

blogger /ˈblɒgə(r)/ *n.* a person who writes for and maintains a blog 博主；写博客的人

influencer /ˈinˌfluənsər/ *n.* ① a person who is able to generate interest in something (such as a consumer product) by posting about it on social media 网红 ②a person who inspires or guides the actions of others 影响者

individual /ˌɪndɪˈvɪdʒuəl/ *n.* a single human being as distinct from a group class or family 个人

impressive /ɪmˈpresɪv/ *a.* evoke admiration through size quality or skill; grand imposing or awesome 令人赞叹的；雄伟的，壮丽的，令人敬畏的

niche /niːʃ; nɪtʃ/ *n.* an opportunity to sell a particular product to a group of people 商机；市场定位

endorse /ɪnˈdɔːs/ *v.* ① to say in an advertisement that you use and like a particular product so that other people will want to buy it（在广告中）宣传，代言（某一产品）② declare one's public approval or support of 公开赞同；支持，认可

present /ˈpreznt/ *v.* ① ~ **sth** ~ **sth/sb/yourself as sth** to show or describe sth/sb in a particular way（以某种方式）展现，显示，表现 ② ~ **sb with sth** | ~ **sth (to sb)** to give sth to sb, especially formally at a ceremony 把 …… 交给；颁发；授予

perspective /pəˈspektɪv/ *n.* ~ **(on sth)** a particular attitude towards sth; a way of thinking about sth 态度；观点；思考方法

adore /əˈdɔː(r)/ *v.* love or like sb or sth very much 热爱；喜爱

buzz /bʌz/ *n.* [sing.] (*informal*) a strong feeling of pleasure, excitement or achievement（愉快、兴奋或有成就感的）强烈情感

recreate /ˌriːkriˈeɪt/ *v.* to make sth that existed in the past exist or seem to exist again 再现；再创造

hippie /ˈhɪpi/ *n.* a usually young person who rejects the way that most people live 嬉皮士

geographic /ˌdʒɪəˈgræfɪk/ *a.* of or relating to geography 地理的；地理学的

trendy /ˈtrendi/ *a.* very fashionable or up to date in style or influence 时髦的

hip /hɪp/ *a.* (*informal*) following the latest fashion, especially in popular music and clothes（尤指流行音乐、服饰）时髦的，新潮的

mighty /ˈmaɪti/ *a.* processing great and impressive power or strength, especially on account of size 强大的；强有力的

imitate /ˈɪmɪteɪt/ *v.* to take or follow as model 模仿；仿效

fanbase /ˈfænbeɪs/ *n.* a group of fans for a particular sport or team 粉丝团

spokesperson /ˈspəʊkspɜːsn/ *n.* ~ **(for sb/sth)** a person who speaks on behalf of a group or an organisation 发言人

icon /ˈaɪkɒn/ *n.* a famous person or thing that people admire and see as a symbol of a particular idea, way of life, etc. 崇拜对象；偶像

splash /splæʃ/	*n.* (*informal*) a striking or exciting effect or event 引人注目的效果，令人激动的事件
everlasting /ˌevəˈlɑːstɪŋ/	*a.* lasting forever or a very long time 永恒的；永久的
take off	to spring into wide use or popularity 广泛兴起或流行
take sth by storm	to win or gain huge and widespread success or popularity very rapidly 使风靡一时；声名鹊起
not necessarily	not definitely or always; possibly not 不一定
look up to	to admire or view one as a role model 尊敬；敬仰
pull off	to be able to perform or complete something, especially in the face of hardships, difficulties, or setbacks 完成；努力实现
go wild	to become very excited or enthusiastic (about sb or sth) 狂热；发狂
turn to sb	to seek or rely on sb or sth as an aid or for assistance through a difficult situation 求助于
find its way to	to arrive at a particular place or situation unintentionally or in a roundabout way 意外或绕行到达特定的地点
up to date	modern; fashionable 现代的；最新的；时髦的；新式的

Reading Comprehension

Understanding the text

Choose the best answer to each of the following questions.

1. Why do people go fanatical for khaki pants?

   A. Because they look extremely fashionable.

   B. Because they are cheap.

   C. Because the pants are really comfortable.

   D. Both A and C.

2. In the less-digital decades of the past, fashion trends used to emerge via ________.

   A. bloggers

   B. runway shows

   C. influencers

   D. fashion houses and magazines

3. Why do various trends emerge from runway looks?

   A. Because runway looks are based on the original and creative ideas of the designers.

   B. Because runway looks are haute couture.

   C. Because runway looks are more fabulous.

   D. Because runway looks are more professional.

4. Why do bloggers and influencers become a major driving force that contributes to the process of trend creation?

   A. Because their tastes are good.

   B. Because online blogs and social media platforms prevail nowadays.

   C. Because they are very fashionable.

   D. Because they lead the trend.

5. Which source shows that anyone can play an impact on fashion trend?

   A. Celebrities.

   B. Street style.

   C. Bloggers and influencers.

   D. Fashion shows.

Mind mapping

According to your understanding, what are the determining factors of fashion trends? Please make an overview of them and present it in a mind map.

Language Enhancement

Words in use

Fill in the blanks with the words given below. Change the form when necessary. Each word can be used only once.

peculiar	surge	fanatical	dramatically	motivate
compile	imminent	anticipation	individual	endorse

1. When communicating with their kids, parents should always maintain their sense of humor and pleasant mood to ________ the kids to continue speaking.
2. Because of Hurricane Ike, large and dangerous waves ________ up to six meters above normal, triggering the shipwreck.
3. Jenny shook her thoughts away, her heart hammering with both ________ of seeing him again and fear.
4. She pronounced the syllables of the name with a ________ clearness, as if she had tapped on two silver bells.
5. She knew that at least this ________ and faithful follower, and his family, would stand behind him to the end.
6. Overexposure to the sun can be the cause of heatstroke, when the body stops sweating and body temperature rises ________.
7. Usually, a celebrity will ________ one of his or her fan sites if they are particularly well done.
8. Our work was like that of geologists who compile the signs of an ________ earthquake or volcanic eruption.

9. He must recognize that his own ________ choices and behaviours must be coordinated with the ongoing activities of his surroundings.
10. All they are allowed to do is ________ statistics from the information provided.

Expressions in use

Fill in the blanks with the expressions given below. Change the form when necessary. Each expression can be used only once.

take by storm	not necessarily	turn to	pull off
look up to	find its way to	up to date	take off

1. Some of that money is expected to ____________ emerging markets, which offer higher yields and prospects for asset-price growth.
2. Twitter has ____________ the web ____________ these past couple of years, and consequently many web developers have tried their hand at making a clone of the popular web service.
3. If they're able to ____________ the merger（合并）, they would form the largest single corporation in the world.
4. I wish you wouldn't ____________ alcohol when you feel stressed about work.
5. They also provide a great way for customers to stay ____________ on the changing content of an app without having to log in repeatedly.
6. Because it is a more traditional style of dancing, we never expected it to ____________ like it has.
7. You have to be pretty popular to get elected, so we conclude that Chinese people in general ____________ and admire their scientists.
8. Income, while being a common primary motivator, is ____________ always the most appropriate mechanism to motivate the sales force.

Sentence structure

I. Complete the following sentences by translating the Chinese into English, using "when it comes to ..." structure.

Model: ____
________.（当谈到引领时尚的最好办法时）, fashion shows are definitely the first to bear the brunt.
→ When it comes to the best ways leading fashion trends, fashion shows are definitely the first to bear the brunt.

1. ____（关于赛车）, there are two things that the designers and engineers keep in mind.
2. I try to be a radical in political and social ways, but ______________________ ____（当谈到技术方面时，我是一个十足的保守派）.
3. I don't take risks ____（在尝试新奇食物方面）, but Jerry tried pretty well everything.

II. Rewrite the following sentences by using "there is little/ no doubt that ... " structure.

Model: It is indisputable that celebrities are style icons as many people turn to them for style and fashion inspiration.
→ There is no doubt that celebrities are style icons as many people turn to them for style and fashion inspiration.

1. It is certain that many people use social media apps to kill time in their day. Yet, when used for good, it can be a powerful productivity tool.
   ____
   ____.
2. It is indisputable that all this new technology is changing the way we work and offering many new ways of working.
   ____
   ____.

3. It is undeniable that the company prefers to employ capable men or hard-working ones.

____

____.

Extensive Reading

The Impact of Minimalism on Modern Fashion

1 It is one of fashion's great ironies that, beyond the spectacle of the runway show, and away from the peacocking gestures captured by street-style snappers, many people working in fashion, those actually there to graft; applying make-up, frantically penning show reports in the back of taxis are often dressed head-to-toe in black. American *Vogue*'s Grace Coddington, for example, might have art directed some of the most visually arresting, colour-drenched shoots the publication has ever seen, but she rarely strays from her signature all-black wardrobe.

2 This minimal approach to fashion could owe itself to the defiant nature of the industry's creatives—after all, isn't it the punkest thing you could do to reject it altogether and just wear one colour when constantly presented with new fashions? It may also have something to do with professionalism, wanting to take a considered back seat, allowing the work itself to grab attention when opposed to claiming it for yourself. Interestingly, this approach bears similarity to the core principles of the minimalist artistic movement, which appeared in New York in the 1950s and 1960s and railed against the over-complication of bright, expressive colours and patterns, preferring instead to employ monochrome black or white as a

palette cleanser, to refocus the mind and open up a new discussion around aesthetics. Here, we explore how some of the key components of the minimalist movement have manifested in the fashion industry...

3 Although minimalist works are now famous in their own right, the actual construction of minimalist sculptures afforded them an anonymous feel, free of ego stamped all over like the works of action painters of the time, who seemingly marked their canvases territorially. Works like Carl Andre's sculptures, which were neither carved, built or modelled, but simply constructed by positioning raw materials like bricks and blocks with no fixatives to hold them in place, were a swerve away from the instantly recognisable works of characterful American painters like Jackson Pollock, Clyfford Still and Hans Hofmann. As Minimalist art pioneer Sol Le Witt said, "Every generation renews itself in its own way; there's always a reaction against whatever is standard."

4 A kindred spirit, both in terms of its minimalist aesthetic and the "invisible" presence of the designer in the work, is the original incarnation of Maison Martin Margiela, founded by designer Martin Margiela, who upheld a sense of anonymity at the house until John Galliano was appointed creative director in 2014. The Margiela aesthetic not only feels in synergy with the minimalist approach—painting everyday items white, and repurposing ordinary household objects like crystal doorknobs—but equally in the way the designer(s) shied away from the limelight, instead letting the work stand for itself, even his models anonymised.

5 The minimalist colour palette is pretty strictly limited to monochrome black or white but, rather than simplifying its themes, taking colour and pattern out of the equation allows the work to explore bigger concerns as form takes centre stage. In fashion, Issey Miyake's famous Bao Bao bags have many occasional, seasonal incarnations of colour variation, but are traditionally white or black, thus the geometric form of interconnected cubes becomes the focal point. Fellow Japanese avant-garde designer Yohji Yamamoto also tends to work only in monochrome, his reasoning, too seeming to be its low-key quality—as he once explained, "Black is modest and arrogant at the same time. Black is lazy and easy—but mysterious. But above all black says this: 'I don't bother you—don't bother me'."

6 The same can be said of plain white. Speaking in *Afterzine* in 2010, Peter Saville (the designer of Joy Division's iconic *Unknown Pleasures* sleeve art and New Order's *Blue Money* amongst myriad other iconic images) mused totally white minimalism design: "These are all responses to the overcrowded visual environment that we exist in. One response to that is to make things bolder and brasher. The other approach is towards the nothing. White space, white furniture which is technical and laboratory-like has an academic intelligence to it. It's the suggestion of negative space as a modern luxury. It's negative space as thinking space. It is a luxury. Time and space are modern luxuries." His words echo almost the exact sentiment of painter Frank Stella, who produced a series of totally black works: "After all, the aim of art is to create space—space that is not compromised by decoration or illustration, space within which the subjects of painting can live."

7 A move towards minimalism wasn't just a shift in visual culture in 1950s and 1960s America; musicians also sought to create new, pared-down sonic spaces. Terry Riley's ambient 1964 composition *In C* is often cited as the first minimalist work in music. The next year, New York musician Steve Reich composed *It's Gonna Rain*, based around recordings of a sermon on the end of the world given by a black Pentecostal street-preacher, created using the favoured minimalist method of deconstructing and resequencing snippets using multiple tape loops. It's a technique that would later be adopted by other experimental artists, one of the first being Brian Eno for *My Life in the Bush of Ghosts*. Fast forward to the present day with artists like Japanese artist/musician Ei Wada of the Open Reel Ensemble, who soundtracked the Issey Miyake S/S16 fashion show, are using similar techniques. Ei Wada's trademark experimental sounds are made with a remodelled tape recorder.

8 Avant-garde fashion's interest in experimental sounds doesn't end with leftfield noise producers, other designers have played with silence and the

ability of their designs to sonically punctuate their live presentations. "For A/W12 womenswear, Comme des Garçons' Rei Kawakubo carpeted her catwalk with velvet and added no sound, the models parading the collection only to the muffled clomps of their shoes, before a swell of electronica announced the finale," wrote Harriet Baker in her report for AnOther. "And at the recent round of S/S15 shows, playing with the traditional MO has become more prominent, with houses tempering their musical soundtracks with noise or silence. Rick Owens turned down the volume on his classical soundtrack to draw attention to the serrated wooden clogs cracking down the marble runway."

Unit 7

Brand Stories

The clothing I prefer is the one I create for a life that does not yet exist, the world of tomorrow.

— *Pierre Cardin*

Clothes could have more meaning and longevity if we think less about owning the latest or cheapest thing and develop more of a relationship with the things we wear.

— *Elizabeth L. Cline*

Pre-Reading Activities

1. Match the slogans with the corresponding Chinese clothes brands.

1 Anything is impossible!	A. PEAK
2 Keep moving!	B. ANTA
3 I can play!	C. LI-NING
4 Dance with wolves.	D. BOSIDENG
5 To be No.1!	E. SEPTWOLVES
6 We warm the world.	F. ERKE

2. Listen to the passage and fill in the missing information.

"Since it was founded, Louis Vuitton has been a byword for stylish, elegant luggage." This comment from its website highlights (1) ____________________ from its very beginnings in 1854. Louis Vuitton worked for the French royal family and (2) ____________________ of what made good luggage. His famous LV monogram is found on most of today's products, ranging from trunks, bags and purses to shoes, watches and sunglasses. The company sells most of its products in (3) ____________________ in up-market department stores. Another cornerstone of the company's success is (4) ____________________. The website says: "The choice of the finest materials, experienced craftsmen and the extreme care devoted to all manufacturing phases of our items, perpetuate and renew our (5) ____________________."

3. Discuss the following questions with your partner.

(1) Which is most important to you in deciding what to buy, brand name, quality or price? And why?

(2) The evolution of fashion has changed people's shopping habits. The arrival of fast fashion makes stylish clothes more affordable. How did fast fashion

happen? What are the common key factors contributing to the success of fast fashion brands?

(3) The fashion industry abounds with prominent figures and fabulous stories that inspire a lot of people to follow their footsteps. Tell the successful fashion story that impresses you most and share your story in class.

ANTA's Evolution & Revitalization

The history of ANTA

1 ANTA, founded in 1991 by Ding Shizhong, is the world's third largest **sportswear** company by market **capitalization**. It has a sixth of the Chinese sportswear market, but revenues have grown by 40% in each of the past two years, double the rate of the industry. In 2018 they sold 60 million pairs of shoes and operating **profits** hit 1.2 billion dollars in 2019. The ANTA Group has over 12,000 stores in China and are mass-market kings. How did ANTA become so **dominant** in the mass-market **segment** and how, using **strategies** ANTA Group will attempt to **carve** themselves a bigger portion of the high-end market?

2 In China, the general feeling about ANTA is that it is cheap. The perception stems from a **setback** the company experienced in 2011. An overly **rapacious expansion** led to the company opening too many stores and **stocking** too many products. To save costs, over 600 stores were closed in one year. But this didn't solve the problem of the **massive backlog** of ordered goods. In order to rid themselves of this stock, ANTA dropped their pricing strategy and entered the low-end market in an attempt to **flog** all remaining stock. They cut the costs and cleared stock, but having fought the price war and coming out the other side, their **foothold** on the footwear

market was weak, which caused the brand long-term **reputational** damage. What had once been a famous **domestic** brand in the eyes of consumers now becomes a market stall.

How ANTA pulled themselves up by Klay Thompson's bootstraps

3 ANTA managed to turn this weakness into a strength. Instead of attempting to leave the **battlefield** of the price war, ANTA decided to change their **tactics**. They signed with very **high-profile** NBA basketballers Rajon Rondo, Luis Scola, Kevin Garnett, and more recently, Klay Thompson to their brand. This would **solidify** their position in the eyes of many Chinese basketball fans, which is still China's most popular sport in terms of interest.

4 Their player-branded basketball shoes are priced in the affordable range at 399 RMB, generally about half the price of their Nike **counterparts**. This was a shock move to many, to have a high-profile NBA player whose own branded shoes would be sold at half the price of those of his teammates. Some also predicted that ANTA was shooting itself in the foot with this strategy. Signature model basketball shoes generally sell between 30,000 and 80,000 pairs a year. However, in the first 2 quarters after the release of KT Fire (Klay Thompson **affiliated**) shoes, sales had surpassed 100,000 pairs.

5 With these celebrities on board ANTA introduced a new slogan "实力无价," which means one can't put a price on somebody's strength and ability. The **concept** is to show that ANTA still offers a level of quality good enough for elite NBA players, but at a lower price, and that ANTA's product quality has always been high. However, one **downside** for the brand is that this marketing **campaign** limits their ability to **sharply** raise prices in the future.

ANTA on the path to becoming mass-market royalty

6 In China, **Consumption** Upgrade refers to the way in which consumer

demands of the middle class have changed since the country opened up. The middle class in China are accepted to be around 200 million strong. ANTA is targeting those below the middle class who have more modest amounts, but still have **expendable** income.

7 The problem that ANTA confronts with this consumer group is in making them feel that what they are buying isn't just "cheap." Customers in China expect to pay around 120–300 RMB for domestically produced sneakers. Carefully, so as not to **garner** a reputation of selling cheap shoes, ANTA sets its price at a slightly higher value within the range while special models are priced higher than expected but within the affordable range. This allows customers to feel they are buying a product in good taste while still getting it at a good price, thus they can distinguish between cheap shoes and the shoes that they are buying.

ANTA fights perception problems in the middle-class market by acquiring competitors

8 ANTA is not seen as cool in the eyes of young Chinese people and is virtually unheard of in the west. However, the company has a very large sports apparel **distribution** network in China, which it can leverage to sell some of the "cool brands" it has acquired in the past, such as FILA in 2009. Other brands under ANTA's name, including Sprandi, Descente and Kolon Sport etc., have grown even faster, achieving gross profit growth of more than 44% each year from 2018 to 2020 and **offset** the loss of the brand ANTA, achieved total profit growth of 10.7%.

9 More recently under a **consortium**, Amer Sports, a Finnish sporting goods company that owns brands such as Wilson and Solomon. These cooler brands will attempt to compete in the higher-end market while ANTA itself can focus on the mass market. ANTA's **sponsorship** of the 2022 Winter Olympics will help market the brand domestically and internationally. To cater to various fast-growing, middle-class markets, they have taken over different high-end brands such as Descente, a Japanese Winter sports brand

to cover the **burgeoning** winter sports market in China.

ANTA still largely relies on offline distribution

10 According to Chen Shaozu, General Manager of ANTA's Fuzhou branch, about 70% of ANTA sales are through brick-and-mortar stores with only 30% of sales happening across online platforms. However, through cooperation with these platforms, ANTA has boosted sales by over 40% year-on-year. During the Tmall Double-11 Shopping Festival, ANTA and FILA ranked third and fifth **respectively** in terms of sales, under the sports and outdoor **category**. And ANTA KIDS ranked third in sales under the kids' clothing and footwear category. The large **reservoir** of data that ANTA holds on its consumers (250 million) will help **reveal** consumer **insights** and will **generate** precision marketing tactics.

What does the future hold in store for ANTA?

11 Fortunately for ANTA's plan, the sports market is not showing any signs of slowing down, and is being **bolstered** by apps like KEEP and gyms like SUPERMONKEY. As part of a successful brand strategy, ANTA have tied down the mass market and despite being recognized as being a cheap brand, they are **simultaneously** perceived to be **decent** in terms of quality. ANTA has done well in **diversifying** its **portfolio** by purchasing **prestige** foreign brands in order to cater to new markets and to high-end consumers.

Notes

Ding Shizhong 丁世忠（1970—，安踏集团创始人丁和木之子，现为安踏集团董事局主席兼 CEO/ 亚玛芬董事会主席）

Klay Thompson 克莱 · 汤普森（1990—，美国职业篮球运动员）

Rajon Rondo 拉简 · 朗多（1986—，美国职业篮球运动员）

Luis Scola 路易斯 · 斯科拉（1980—，阿根廷职业篮球运动员）

Kevin Garnett 凯文 · 加内特（1976—，美国职业篮球运动员，绰号“狼王”）

FILA 斐乐（意大利百年运动时尚品牌，始创于 1911 年）

Sprandi 斯潘迪（英国的运动时尚鞋品牌，创始人为 Dinesh Bindal）

Descente 迪桑特（日本高端专业运动品牌，始创于 1935 年）

Kolon Sport 可隆户外（韩国顶尖专业户外品牌，始创于 1973 年）

Amer Sports 亚玛芬体育（世界顶级体育器材品牌管理集团公司，于 1959 年创立于芬兰）

Wilson 威尔胜（世界运动产品的焦点和运动器材制造业的领导者之一）

Solomon 所罗门（法国全球顶级户外运动品牌，始创于 1947 年）

New words and phrases

revitalization /ˌriːvaitəlaiˈzeiʃn/ *n.* [U] the act of making someone or something have strength, energy or health again 复兴、复苏

sportswear /ˈspɔːtsweə(r)/ *n.* [U] clothes that are worn for playing sports, or in informal situations 运动装；休闲服

capitalization /ˌkæpɪtəlaɪˈzeɪʃn/ *n.* [U] ① the total value of a company's shares 资本总额 ② the total value of all the shares on a stock market at a particular time 市值；市场资本总额

profit /ˈprɒfɪt/ *n.* ① [C, U] ~ **on sth** ~ **from sth** the money that you make in business or by selling things, especially after paying the costs involved 利润；收益；赢利 ② [U] (*formal*) the advantage that you get from doing sth 好处；利益；裨益

dominant /ˈdɒmɪnənt/	*a.* ① more important, powerful or noticeable than other things 首要的；占支配地位的；占优势的；显著的 ② (*biology*) A dominant gene causes a person to have a particular physical characteristic, for example brown eyes, even if only one of their parents has passed on this gene.（基因）显性的；优势的
segment /ˈsegmənt/	*n.* [C] a part of sth that is separate from the other parts or can be considered separately 部分；份；片；段
strategy /ˈstrætədʒi/	*n.* ① [C] ~ **(for doing sth)** ~ **(to do sth)** a plan that is intended to achieve a particular purpose 策略；计策；行动计划 ② [U] the process of planning sth or putting a plan into operation in a skillful way 策划；规划；部署；统筹安排
carve /kɑːv/	*v.* ① to make objects, patterns, etc. by cutting away material from wood or stone 雕刻 ② [*no passive*] ~ **sth (out) (for yourself)** to work hard in order to have a successful career, reputation, etc. 艰苦创业；奋斗取得（事业、名声等）
setback /ˈsetbæk/	*n.* [C] a difficulty or problem that delays or prevents sth, or makes a situation worse 挫折；阻碍
rapacious /rəˈpeɪʃəs/	*a.* (*formal*) wanting more money or goods than you need or have a right to 贪婪的；贪欲的；强取的
expansion /ɪkˈspænʃn/	*n.* [U, C] an act of increasing or making sth increase in size, amount or importance 扩张；扩展；扩大；膨胀

stock /stɒk/	*v.* ① (of a shop/store) to keep a supply of a particular type of goods to sell 存货 ② [often passive] ~ **sth (with sth)** to fill sth with food, books, etc. 贮备，贮存（食物、书籍等）
massive /ˈmæsɪv/	*a.* ① very large, heavy and solid 巨大的；大而重的；结实的 ② extremely large or serious 巨大的；非常严重的
backlog /ˈbæklɒg/	*n.* [U] a quantity of work that should have been done already, but has not yet been done 积压的工作
flog /flɒg/	*v.* [U] ① (*BrE, informal*) ~ **sth (to sb)** ~ **sth (off)** to sell sth to sb 出售（某物给某人） ② [often passive] to punish sb by hitting them many times with a whip or stick 鞭笞，棒打（作为惩罚）
foothold /ˈfʊthəʊld/	*n.* [C] ① a crack, hole or branch where your foot can be safely supported when climbing 立足处(攀登时足可踩的缝、洞、树枝等） ② (usually sing.) a strong position in a business, profession, etc. from which sb can make progress and achieve success（可以此发展或取得成功的）稳固地位，立足点
reputational /ˌrepjəˈteɪʃənəl/	*a.* of or relating to one's reputation 名声的；声誉的
domestic /dəˈmestɪk/	*a.* ① [usually before noun] of or inside a particular country; not foreign or international 本国的；国内的 ② [only before noun] used in the home; connected with the home or family 家用的；家庭的；家务的

battlefield /ˈbætlfiːld/ *n.* [C] ① a place where a battle is being fought or has been fought 战场 ② a subject that people feel strongly about and argue about 争论主题；斗争领域

tactic /ˈtæktɪk/ *n.* [C] [usually pl.] the particular method you use to achieve sth 策略；手段；招数

high-profile /ˌhaɪˈprəʊfaɪl/ *a.* [usually before noun] attracting a lot of public attention, usually deliberately 高调的；备受瞩目的；知名度高的

solidify /səˈlɪdɪfaɪ/ *v.* ① to become solid; to make sth solid（使）凝固，变硬，变得结实 ② (*formal*) (of ideas, etc.) to become or to make sth become more definite and less likely to change（使）变得坚定，变得稳固，巩固

counterpart /ˈkaʊntəpɑːt/ *n.* [C] a person or thing that has the same position or function as sb/sth else in a different place or situation 职位（或作用）相当的人；对应的事物

affiliated /əˈfɪlieɪtɪd/ *a.* (only before noun) closely connected to or controlled by a group or an organisation 隶属的

concept /ˈkɒnsept/ *n.* [C] ~ **(of sth)** ~ **(that ...)** an idea or a principle that is connected with sth abstract 概念；观念

downside /ˈdaʊnsaɪd/ *n.* [C] (sing.) the disadvantages or less positive aspects of sth 缺点；不利方面

campaign /kæmˈpeɪn/ — *n.* [C] ① a series of actions intended to achieve a particular result relating to politics or business, or a social improvement（有计划的）活动；运动 ② a series of battles, attacks etc. intended to achieve a particular result in a war 一系列军事行动；战役

sharply /ˈʃɑːpli/ — *ad.* ① suddenly and by a large amount 急剧地；突然大幅度地 ② in a critical, rough or severe way 尖刻地；严厉地；猛烈地 ③ in a way that clearly shows the differences between two things 鲜明地；明显地

royalty /ˈrɔɪəlti/ — *n.* [U] members of a royal family 王室成员

consumption /kənˈsʌmpʃn/ — *n.* [U] ① the act of buying and using products 消费 ② the act of using energy, food or materials; the amount used（能量、食物或材料的）消耗，消耗量

expendable /ɪkˈspendəbl/ — *a.* If you regard someone or something as expendable, you think it is acceptable to get rid of them, abandon them, or allow them to be destroyed when they are no longer needed. 可消耗的

garner /ˈgɑːnə(r)/ — *v.* (*formal*) to obtain or collect sth such as information, support, etc. 获得，得到；收集（信息、支持等）

distribution /ˌdɪstrɪˈbjuːʃn/ — *n.* ① [U] (business) the system of transporting and delivering goods（商品）运销，经销，分销 ② [U, C] the way that sth is shared or exists over a particular area or among a particular group of people 分配；分布 ③ [U] the act of giving or delivering sth to a number of people 分发；分送

offset /ˈɒfset/ *v.* ~ **sth (against sth)** to use one cost, payment or situation in order to cancel or reduce the effect of another 抵消；弥补；补偿

consortium /kənˈsɔːtiəm/ *n.* [C] (*pl.* **consortiums** or **consortia**) a group of people, countries, companies, etc. who are working together on a particular project（合作进行某项工程的）财团，银团；联营企业

sponsorship /ˈspɒnsəʃɪp/ *n.* ① [U] the act of sponsoring sb/sth or being sponsored 资助；主办；倡议 ② [U, C] financial support from a sponsor 资助；赞助款

burgeoning /ˈbɜːdʒənɪŋ/ *a.* growing or developing rapidly 迅速发展的，快速生长的；繁荣的

respectively /rɪˈspektɪvli/ *ad.* in the same order as the people or things already mentioned 分别；各自；顺序为；依次为

category /ˈkætəgəri/ *n.* [C] a group of people or things with particular features in common（人或事物的）种类；范畴

reservoir /ˈrezəvwɑː(r)/ *n.* [C] ① (*formal*) a large amount of sth that is available to be used（大量的）储备，储藏 ② a natural or artificial lake where water is stored before it is taken by pipes to houses, etc. 水库；蓄水池

reveal /rɪˈviːl/ *v.* ① ~ **sth (to sb)** to make sth known to sb 揭示；显示；透露 ② to show sth that previously could not be seen 显出；露出；展示

insight /ˈɪnsaɪt/ *n.* ① [U] (*approving*) the ability to see and understand the truth about people or situations 洞察力；领悟 ② [C, U] ~ **(into sth)** an understanding of what sth is like 洞悉；了解

generate /ˈdʒenəreɪt/ *v.* to produce or create sth 产生；引起

bolster /ˈbəʊlstə(r)/ *v.* ~ **sth (up)** to improve sth or make it stronger 改善；加强

simultaneously /ˌsɪmlˈteɪniəsli/ *ad.* in a way that things happen at exactly the same time 同时的

decent /ˈdiːsnt/ *a.* ① of a good enough standard or quality 像样的；相当不错的；尚好的 ② (of people or behaviour) honest and fair; treating people with respect 正派的；公平的；合乎礼节的 ③ acceptable to people in a particular situation 得体的；合宜的；适当的

diversify /daɪˈvɜːsɪfaɪ/ *v.* ① ~ **(sth) (into sth)** (esp. of a business or company) to develop a wider range of products, interests, skills, etc. in order to be more successful or reduce risk 增加 …… 的品种；从事多种经营；扩大业务范围 ② to change or to make sth change so that there is greater variety（使）多样化，变化，不同

portfolio /pɔːtˈfəʊliəʊ/ *n.* [C] (*pl.* **-os**) the range of products or services offered by a particular company or organisation（公司或机构提供的）系列产品，系列服务

prestige /preˈstiːʒ/	*a.* ① [only before noun] admired and respected because it looks important and expensive 名贵的；贵重的；讲究派头的 ② that brings respect and admiration; important 令人敬仰的；受尊重的；重要的 *n.* [U] the respect and admiration that sb/sth has because of their social position, or what they have done 威信；声望；威望
stem from	(*not used in the progressive tenses*) to be the result of sth 是……的结果；起源于；根源是
lead to	to have sth as a result 导致，造成（后果）
rid oneself of sth	to make oneself free from sb/sth that is annoying or causing problems 摆脱；从……中解脱
in an attempt to do sth	an act of trying to do sth, especially sth difficult, often with no success 企图；试图；尝试
in the eyes of sb	according to a particular person or group 在某人的观点中；在某人看来
pull oneself up by your (own) bootstraps	to improve your situation yourself, without help from other people 自力更生
turn … into …	to make sb/sth become sth 使（从……）变成
in terms of sth	used to show what aspect of a subject you are talking about or how you are thinking about it 谈及；就……而言；在……方面
shoot oneself in the foot	to say or do sth stupid that will cause a lot of trouble to himself/herself 搬起石头砸自己的脚；自讨苦吃

refer to sb/sth	① to describe or be connected to sb/sth 描述；涉及；与 …… 相关 ② to look at sth or ask a person for information 查阅；参考；征询
confront sb with sb/sth	to make sb face or deal with an unpleasant or difficult person or situation 使面对，使面临，使对付（令人不快或难处的人、场合）
distinguish between A and/from B	to recognize the difference between two people or things 区分；辨别；分清
cater to sb/sth	to provide the things that a particular type or person wants, especially things that you do not approve of 满足需要；迎合
rely on/upon sb/sth	① to need or depend on sb/sth 依赖；依靠 ② to trust or have faith in sb/sth 信任；信赖
tie sb down (to sth/to doing sth)	to restrict sb's freedom, for example by making them accept particular conditions or by keeping them busy 限制；束缚；牵制

Reading Comprehension

Understanding the text

Answer the following questions.

1. Where did the perception of ANTA being cheap come from?
2. How did ANTA solve the problem of the massive backlog of ordered goods?
3. How did ANTA pull themselves up from the long-term reputational damage?
4. What does the new slogan ANTA introduced mean?
5. Who are the targeted consumers of ANTA?
6. What did ANTA do to cater to various fast-growing middle-class markets?
7. According to Chen Shaozu, what were the sales achieved through brick-and-mortar stores?

8. Did ANTA get success through online platforms?

Critical thinking

Work in pairs and discuss the following questions.

1. According to your understanding, why is it important to maintain and bolster the image of a brand?
2. Is the acquisition strategy really good for ANTA's brand image? And why?

Research project

Keep moving is another advertising slogan of ANTA. This slogan is also used to describe the enterprise spirit of ANTA company. Get more detailed information about the evolution of ANTA and present your ideas on how ANTA keeps moving.

Language Enhancement

Words in use

Fill in the blanks with the words given below. Change the form when necessary. Each word can be used only once.

domestic	dominant	expansion	insight	prestige
reveal	setback	diversify	strategy	category

1. Music shops should arrange their recordings in simple alphabetical order, rather than by ________.
2. Although the ecosystem concept was very popular in the 1950s and 1960s, it is no longer the ________ paradigm.
3. The company plans to sponsor television programs as part of its marketing ________.
4. The Democrats are trying desperately to change the subject to the home front, to talk about ________ issues.
5. The ________ of western capitalism incorporated the Third World into an exploitative world system.

6. A survey of the American diet has ________ that a growing number of people are overweight.
7. He was a man of forceful character, with considerable ________ and diplomatic skills.
8. The islands' economy had received a severe ________ from the effects of hurricane Hugo.
9. Revlon launched a comprehensive marketing program for its ________ brands.
10. Why has language ________ from its ancient origins into more than 5,000 mutually unintelligible varieties?

Banked cloze

Fill in the blanks by selecting suitable words from the word bank. You may not use any of the words more than once.

A. flexibility	F. luxury	K. preferences
B. purchases	G. moderate	L. possessions
C. expensive	H. domestic	M. popularity
D. traditional	I. status	N. progressed
E. witnessed	J. prestige	O. old-fashioned

Bustling Xidan Street in downtown Beijing is home to dozens of clothing stores, featuring both foreign and 1. ________ labels, which is also testimony to the growing 2. ________of domestic brands.

Since the late 1970s, China's garment industry 3. ________ rapid development under China's economic reforms. In 1992, the initiatives to establish a market-oriented economy improved the 4. ________ of Chinese apparel industry. Over time, the Chinese industry has 5. ________ from being largely volume-driven in the 1960s and 1970s to being export-driven in the 1980s and early 1990s to being more consumer-oriented today. In 1995, China became the largest apparel industry in the world. Chinese garments are well-known in Western countries for their 6. ________ or low prices. However, Western luxury brands remain very much a

7. ________ symbol in China. After decades of effort, a few made-in-China brands now produce high quality clothing for high-fashion consumers. Chinese shoppers not only line up outside Chanel and Louis Vuitton stores, they have also started to show off their 8. ________ at NETIGER and Shanghai Tang.

As China is expected to be the world's biggest luxury market soon, Chinese culture and 9. ________ cannot be ignored. In the West, Chinese culture, Chinese designs and China elements are increasingly accepted. In addition, more Chinese are seeking out goods that emphasise their culture. Chinese luxury companies do possess a great opportunity to win the hearts and minds of Chinese consumers, especially when 10. ________ Chinese elements are smartly produced with modern technology and marketing.

Expressions in use

Fill in the blanks with the expressions given below. Change the form when necessary. Each expression can be used only once.

in an attempt to	stem from	rid oneself of	tie … down …
distinguish between	cater to	refer to	rely on

1. Yet the report shows how poor countries' governments often ____________ their own people ______________ in a thicket of useless regulation.
2. The term "time machine", coined by Wells, is now universally used to ____________ a vehicle transporting people into the far future.
3. The administration is just stonewalling ________________ hide their political embarrassment.
4. Why couldn't he ever ____________ those thoughts, those worries?
5. According to the survey, lack of enthusiasm could ____________ concerns about privacy and security.
6. The fundamental problem lies in their inability to ____________ reality and invention.
7. To ____________ customers' desires for popular destinations, the four new

routes are aimed at helping guests find new and exciting locations.

8. Schools are increasingly ____________ money from parents to provide decent stocks of books.

Translation

I. Translate the following paragraph into Chinese.

Li Ning Company is a sports goods company founded in 1990 by Mr. Li Ning, a well-known "gymnastic prince" in China. After decades of exploration, Li Ning has gradually become a leading international sports brand company representing China. Since its inception, it took the lead in establishing a franchise monopoly marketing system throughout the country to sponsor Chinese sports delegations to participate in various domestic and international competitions. Taking innovation as the foundation of brand development, Li Ning has attached great importance to original design. As a leader in the domestic sporting goods industry, Li Ning has actively assumed the social responsibilities of corporate citizens while supporting its own development, aiding disaster-stricken areas, caring for AIDS orphans, and long-term support for sports education in poverty-stricken areas. As a sporting goods company, Li Ning Company intends to become an internationally competitive sporting goods brand and is committed to the creation of professional sports goods with the desire and strength to achieve breakthroughs, and strives to make sports change life and pursue higher goals. The company's focuses on the customer, brand message, retail experience, data, and operation are key ingredients for a successful apparel company, giving them a strong foundation for future growth.

II. Translate the following paragraph into English.

“快时尚”市场在中国发展迅猛。在“快时尚”市场上，国际零售商们出售的服饰与大牌设计师设计的款式相仿，且价格亲民，旨在将最新的时尚趋势尽快带入大众市场。“快时尚”的迅速发展及中国人可支配收入的增加，导致了青少年时尚达人 (fashionistas) 数量的攀升。

Paragraph Writing

How to Elaborate a Paragraph—Unity and Coherence

Effective writing depends on more than just the grammatically correct composition of each sentence. The reader must be able to move easily from one sentence to the next. The sentences should therefore express a coherent train of thought. Together, they must constitute a unified whole in a coherent manner.

Unity

Paragraph **unity** refers to the harmony between the topic sentence and supporting sentences, which means that the entire paragraph should focus on one single idea. The supporting details should explain the main idea. The concluding sentence should end the paragraph with the same idea. A unified paragraph must follow the idea mentioned in the topic sentence and must not deviate from it. Otherwise, the intrusion of irrelevant information can disrupt our understanding of a paragraph.

Ways to preserve unity:

- One way to preserve unity in a paragraph is to start with a topic sentence that shows the main idea of the paragraph. Then, make sure each sentence in the paragraph relates to that main idea. If you find a sentence that goes off track, perhaps you need to start a separate paragraph to write more about that different idea. Each paragraph should generally have only one main idea.

- Another way to think about unity in a paragraph is to imagine your family tree. Draw a quick sketch of your family tree in your notebook. If you were writing an essay about your family, you might write a paragraph about close family members first. Next, you might branch out into another paragraph to write about more distant relatives. Each paragraph would have just one main idea (immediate family, more distant relatives, close family friends), and every sentence in each paragraph would relate to that main idea.

Coherence

Coherence means establishing a relationship between the ideas presented in a paragraph. A coherent paragraph has sentences that all logically follow each other; they are not isolated thoughts. When a text is unified and coherent, the reader can easily understand the main points.

Coherence can be achieved in several ways:

- Using transitions helps connect ideas from one sentence to the next and make your writing much clearer.
- Using pronouns to refer to antecedents is a useful device to make your writing coherent.
- Repeating key nouns or ideas helps the reader remember the main ideas in an essay.
- Structuring each paragraph according to one of the following patterns helps to organise sentences: general to particular; particular to general; whole to parts; question to answer; or effect to cause.

Transitional words and phrases guide readers from one sentence to the next. Although they most often appear at the beginning of a sentence, they may also show up after the subject.

Here is a chart of the transitional devices accompanied with a simplified definition of function:

addition	again, also, and, and then, besides, equally important, finally, first, further, furthermore, in addition, in the first place, last, moreover, next, second, still, too
comparison	also, in the same way, likewise, similarly
concession	granted, naturally, of course
contrast	although, and yet, at the same time, but at the same time, despite that, even so, even though, for all that, however, in contrast, in spite of, instead, nevertheless, notwithstanding, on the contrary, on the other hand, otherwise, regardless, still, though, yet
emphasis	certainly, indeed, in fact, of course

example or illustration	after all, as an illustration, even, for example, for instance, in conclusion, indeed, in fact, in other words, in short, it is true, of course, namely, specifically, that is, to illustrate, thus, truly
summary	all in all, altogether, as has been said, finally, in brief, in conclusion, in other words, in particular, in short, in simpler terms, in summary, on the whole, that is, therefore, to put it differently, to summarize
time sequence	after a while, afterward, again, also, and then, as long as, at last, at length, at that time, before, besides, earlier, eventually, finally, formerly, further, furthermore, in addition, in the first place, in the past, last, lately, meanwhile, moreover, next, now, presently, second, shortly, simultaneously, since, so far, soon, still, subsequently, then, thereafter, too, until, until now, when

Read the following paragraphs. Is there a topic sentence? If so, do all of the other sentences relate to the topic sentence? Can you find any sentences that don't relate?

• The planned community of Columbia, Maryland, was designed as a city open to all, regardless of race, level of income, or religion. When Columbia began in 1967, many cities in the U.S. did not allow people of certain races to rent or buy homes. Its developer, James W. Rouse, wanted to build a new city that had fair and open housing options for everyone. HCC has a building named for James W. Rouse. Today, the city's nearly 100,000 remain diverse, as shown by recent census data.

• College can be expensive and difficult. Critical thinking is a very important skill for college students to develop so that they can be successful in their careers. Employers look for graduates who can understand information, analyse data, and solve problems. They also want employees who can think creatively and communicate their ideas clearly. College students need to practice these skills in all of their classes so that they can demonstrate their abilities to potential employers.

• Bananas are one of Americans' favorite types of fruit. The Cavendish variety, grown in Central and South America, is the most commonly sold here in the U.S. Recent problems with a fungus called Panama disease (or TR4), however, have led to a shortage of Cavendish bananas. Similar problems occurred a few years ago in parts of Asia and the Middle East. Because the fungus kills the crop and contaminates the soil, scientists are concerned that the popular Cavendish banana could be completely eradicated. Bananas contain many nutrients, including potassium and Vitamin B6.

Write an outline for an essay on one of the following topics, edit your outline for unity and coherence, then write the essay.

• The importance of change
• The importance of physical exercise on campus
• The importance of being optimistic

Urban Revivo—Leading the Era of New Retail

1 Urban Revivo is a trend-leading fashion brand, with an innovative, contemporary and **youthful** mood at the center of its brand identity. Established in 2006 by Li Mingguang, who was inspired by a visit to a Zara store in Japan, Urban Revivo has become a $450 million-a-year brand that **claims** to **launch** more than 20,000 new products **annually**.

2 Created to **challenge** the traditional model of fast-fashion production, Urban Revivo has **positioned** itself within the retail sector as a pioneer of **affordable** yet trend-driven high-quality products. The business has always been an unconventional thinker with the fast **luxury** retail sector, embedding new season design and innovation into the **core** of its brand DNA. The brand endeavors to share with its new international customers the most fun, trend-led cross-boundary product experienced within both its luxury retail environments and **digitally** on its modern online **platforms**. UR **represents** a mid-range, slightly boutique, middle-class brand, somewhat like Mango, associated with the growing class of young professionals the founder initially meant to **target**.

3 Throughout the past decade the brand has grown **rapidly** with an increasing number of stores within the Asia-Pacific region and across the world in this space of time, with future exciting plans to **expand** in many more regions across Europe, North America and Japan.

4 Seeking to create an "international fashion **empire**" by focusing on quality fabrics, good cuts, and up-to-date styles inspired by European design, UR has established a solid brand foundation, including a fixed consumer base, strong product **competitiveness** and a leading level of **flexibility** within its supply chain. In 2018, UR made an initial **foray** into London, opening its European **flagship** store in Westfield London. As well as being the largest UR store in London, it also includes an **upgraded** art space to bring local consumers a multi-cultural store experience. The design of the store closely combines artistic play with the **elements** of fashion life and expresses the brand's **vivid** and **diverse** image. London was chosen for the company's European flagship because it is a **landmark** of international fashion and is one of the most **mature** markets in the world, which will allow the brand to examine itself within a key international market, so that they can **continuously** improve the competitiveness of its products and brand image.

5 According to Leo Li, Company Brand Founder and CEO, UR will pay

more attention to product development and improvement, and continuously improve the quality of the store's operations. In addition, the development of online channels is also a key strategy that will enable customers to recognize and understand the brand. At the same time, the company is also considering cooperating with more European e-commerce platforms, so consumers can get closer to the brand. Exploration of potential new markets is an important factor for the brand moving forward. **Despite** this, they still see its home market of China as having great potential, along with the Asia-Pacific region and the US market. Consumers' perceptions of fashion are different, but the demand for fashion is universal.

6 As a fashion brand, Urban Revivo does not change their offer depending on the market they are operating in. Instead, they **capture** the most **cutting-edge** fashion trends, combined with trend analysis, to provide its customers with the most fashionable new seasons and let consumers **acquire** and **perceive** the brand. "Urban Revivo has found a balance between fashion and luxury. We not only have the most fashionable products; we also have the highest quality products at a **reasonable** price. We also attach more importance to the creation of brand aesthetics, hoping to give consumers a richer and **well-rounded** experience. I believe this is why we are deeply loved by our customers," says Leo.

7 The company always insists that products are the core of the brand, so a lot of time is spent on product creation, including not only the **gradual** creation of the product, but the extremely flexible supply chain, **efficient** logistics and **in-depth** exploration and pursuit of the perfect consumer experience to enhance the fashion expression that Urban Revivo brings to its consumers.

8 Since its establishment in 2006, the business has made rapid progress, becoming the creator of the world's first fast luxury fashion business model. With "fashion UR, technology UR" as the **strategic** core, UR adheres to fashion at its root but is technology-based. UR continues to increase investment in product design and technology, create the latest fashion trend products for customers, and use technology to **enable** management to reduce costs, make more **efficient** and more accurate decisions, realize the **transformation** from traditional enterprises to innovative intelligent business enterprises, and realize the dream of making the world more fashionable. UR's **mission** is to improve the quality of people's fashion life, and its goal is to become the world's largest fashion group and China's first transformed fashion brand.

9 "Play fashion is the core of our brand, and it is this **essence** that has always been adhered to. Together with consumers, we are looking for **freshness** and attitude in our fashion offerings," explains Leo. Leo feels that the challenges **confronting** the company are its continued development and expansion into a global market that is **constantly** changing, and to continue to innovate and to ensure that effective innovation is always at the forefront of Urban Revivo's thinking and the driving force for progress.

10 Today is the era of new retail, therefore Urban Revivo **aspires** to strengthen the construction of new retail and **omni-channel** to become the world's leading fashion smart business enterprise.

Notes

Urban Revivo 快尚时装(简称 UR)

New words and phrases

era /ˈɪərə/ *n.* [C] a period of time of which particular events or stages of development are typical 时代；年代；纪元

retail /ˈriːteɪl/ *n.* [U] the activity of selling goods to the public, usually in shops or stores 零售，零卖

youthful /ˈjuːθfl/ *a.* having the qualities that are typical of young people 年轻人特有的；富有青春活力的，朝气蓬勃的

claim /kleɪm/

v. ① to say that sth is true or is a fact, although you cannot prove it and other people might not believe it 宣称；声称；断言 ② to ask for sth of value because you think it belongs to you or because you think you have a right to it 要求（拥有）；索取；认领 ③ to make a written demand for money from a government or organisation because you think you have a right to it 向政府或组织）索取（钱）；索（款）；索赔

launch /lɔːntʃ/

v. ① to make a product available to the public for the first time （首次）上市；发行；推出 ② send (a missile, satellite or spacecraft) on its course or into orbit 发射（人造卫星、导弹或航天器）

n. [U] making a start in a career or vocation 开始；创办

annually /ˈænjuəli/

ad. once a year 一年一次地，每年地

challenge /ˈtʃæləndʒ/

v. ① to invite someone to compete or take part, especially in a game or argument（尤指比赛或辩论）挑战；邀请 ② to question if sth is true or legal 质疑，怀疑

n. [C, U] (the situation of being faced with) sth that needs great mental or physical effort in order to be done successfully and therefore tests a person's ability 挑战；难题；考验

position /pəˈzɪʃn/ *v.* ① promote (a product, service, or business) within a particular sector of a market, or as the fulfilment of that sector's specific requirements 为（产品，业务）打开销路；确立（产品，服务，业务的）行业地位 ② to put sb/sth in a particular position 安装；安置；使处于
n. the place where sth or sb is, often in relation to other things 位置，方位，地点

affordable /əˈfɔːdəbəl/ *a.* not expensive 付得起的；买得起的

luxury /ˈlʌkʃəri/ *n.* ① [C] sth expensive that is pleasant to have but is not necessary 奢侈品，名贵品 ② [U] great comfort, especially as provided by expensive and beautiful things 豪华，奢华

core /kɔː(r)/ *n.* ① [sing.] the basic and most important part of sth 核心，关键，最重要的部分 ② [C] the hard central part of some fruits, such as apples, that contains the seeds（水果的）核，心

digitally /ˈdɪdʒɪtəli/ *ad.* in a way that uses or relates to digital signals and computer technology 以数字方式；数码地

platform /ˈplætfɔːm/ *n.* [C] ① the type of computer system or the software that is used 计算机平台 ② a flat raised structure, usually made of wood, that people stand on when they make speeches or give a performance 讲台；舞台 ③ a long, flat raised structure at a railway station, where people get on and off trains (火车或地铁站的) 站台，月台

represent /ˌreprɪˈzent/ *v.* ① to be a member of a group and act or speak on their behalf at an event, a meeting, etc. 代表 ② to act or speak officially for sb and defend their interests 作为……的代言人；维护……的利益

target /ˈtɑːɡɪt/ *v.* ① to direct advertising, criticism, or a product at someone 面向，把……对准（某群体）② to aim an attack, or a bullet, bomb, etc., at a particular object, place, or person 把…作为攻击目标

n. [C] ① **~ (for sb/sth) ~ (of sth)** an object, a person or a place that people aim at when attacking （攻击的）目标，对象 ② an object that people practise shooting at, especially a round board with circles on it 靶；靶子

rapidly /ˈræpɪdli/ *ad.* in a fast or sudden way 迅速地；很快地

expand /ɪkˈspænd/ *v.* ① to increase in size, number, or importance, or to make sth increase in this way 扩大；增加，增强（尺码、数量或重要性）② If a business **expands** or is **expanded**, new branches are opened, it makes more money, etc. 扩展，发展（业务）

empire /ˈempaɪə(r)/ *n.* [C] ① a group of countries ruled by a single person, government, or country 帝国 ② a group of commercial organisations controlled by one person or company 大企业；企业集团

competitiveness /kəmˈpetətɪvnəs/ *n.* [U] the fact of being able to compete successfully with other companies, countries, organisations, etc. 竞争力

flexibility /ˌfleksəˈbɪləti/ *n.* [U] ① the ability to change or be changed easily according to the situation 灵活性；适应性 ② the ability to bend or to be bent easily without breaking 弹性；柔韧性

foray /ˈfɒreɪ/ *n.* [C] ~ **(into sth)** an attempt to become involved in a different activity or profession（改变职业、活动的）尝试

flagship /ˈflæɡʃɪp/ *n.* [C] ① the best or most important product, idea, building, etc. that an organisation owns or produces 王牌产品；最佳服务项目；主建筑物 ② the main ship in a fleet of ships in the navy 旗舰

upgrade /ˌʌpˈgreɪd/ *v.* ① to improve the condition of a building, etc. in order to provide a better service 提高（设施、服务等的）档次；改善；使升格 ② to make a piece of machinery, computer system, etc. more powerful and efficient 使(机器、计算机系统等)升级；提高；改进 ③ ~ **sb (to sth)** to give sb a more important job 提升；提拔

element /ˈelɪmənt/ *n.* [C] ① ~ **(in/of sth)** an important quality or feature that a situation, activity or process has or needs 要素；特点 ② a part of sth 组成部分

vivid /ˈvɪvɪd/ *a.* ① (of descriptions or memories) producing clear, powerful, and detailed images in the mind（记忆和描写）清晰的；生动逼真的 ② very brightly coloured（颜色）鲜艳的

diverse /daɪˈvɜːs/ *a.* ① including many different types of people or things 各种各样的 ② very different from each other 不同的

landmark /ˈlændmɑːk/ *n.* [C] ① a building or feature which is easily noticed and can be used to judge your position or the position of other buildings or features 地标，陆标 ② an important stage in sth's development 里程碑

mature /məˈtʃʊə(r)/ *a.* ① denoting an economy, industry, or market that has developmed to a point where substantial expansion and investment no longer takes place（经济，行业，市场）成熟的；发展余地不大的② behaving in a sensible way, like an adult 明白事理的；成熟的；成人似的

continuously /kənˈtɪnjuəsli/ *ad.* without a pause or interruption 连续不断地

despite /dɪˈspaɪt/ *prep.* used to show that sth happened or is true although sth else might have happened to prevent it 即使；尽管

capture /ˈkæptʃə(r)/ *v.* ① to succeed in getting control of sth that other people are also trying to control 夺得；赢得；争得 ② to catch a person or an animal and keep them as a prisoner or in a confined space 俘虏；俘获；捕获

cutting-edge /ˌkʌtɪŋ ˈedʒ/ *a.* the movement of large numbers of people, birds or animals from one place to another 领先的，最新的；先进的，尖端的
n. [sing.] the most modern stage of development in a particular type of work or activity 前沿

acquire /əˈkwaɪə/ — *v.* ① If you acquire something, you buy or obtain it for yourself, or someone gives it to you. 获得 ② If you acquire something such as a skill or a habit, you learn it, or develop it through your daily life or experience. 习得

perceive /pəˈsiːv/ — *v.* to notice or become aware of sth 感知到；注意到；意识到；察觉到

reasonable /ˈriːznəbl/ — *a.* ① (of price) not too expensive 不太贵的；公道的 ② ~ **(to do sth)** fair, practical and sensible 公平的；合理的；有理由的；明智的

well-rounded /ˌwel ˈraʊndɪd/ — *a.* involving or having experience in a wide range of ideas or activities or providing knowledge in a number of different areas 全面的

gradual /ˈɡrædjʊəl/ — *a.* changing or developing slowly or by small degrees 逐渐的；逐步的

efficient /ɪˈfɪʃnt/ — *a.* doing sth well and thoroughly with no waste of time, money, or energy 效率高的；有能力的

in-depth /ˌɪn ˈdepθ/ — *a.* done carefully and in great detail 彻底的；深入详尽的

strategic /strəˈtiːdʒɪk/ — *a.* relating to the most important, general aspects of sth such as a military operation or political policy, especially when these are decided in advance 战略上的

enable /ɪˈneɪbl/ — *v.* ① to make it possible for sth to happen or exist by creating the necessary conditions 使成为可能；使可行；使实现 ② to make it possible for sb to do sth 使能够；使有机会

transformation /ˌtrænsfəˈmeɪʃn/ *n.* [C, U] ~ **(from sth) (to/into sth)** a complete change in sb/sth（彻底的）变化，改观；转变；改革

mission /ˈmɪʃn/ *n.* [C] the result that a company or an organisation is trying to achieve through its plans or actions 使命

essence /ˈesns/ *n.* [U] ~ **(of sth)** the most important quality or feature of sth, that makes it what it is 本质；实质；精髓

freshness /ˈfreʃnəs/ *n.* [U] the quality of being new or original, and usually therefore interesting 新鲜感，新鲜度

confront /kənˈfrʌnt/ *v.* ①（of problems or a difficult situation）to appear and need to be dealt with by sb 面临（问题、困难等），使……无法回避 ② to face sb so that they cannot avoid seeing and hearing you, especially in an unfriendly or dangerous situation 面对；对抗；与（某人）对峙

constantly /ˈkɒnstəntli/ *ad.* all the time or repeatedly 始终；一直；重复不断地

aspire /əˈspaɪə(r)/ *v.* ~ **(to sth/to do sth)** to have a strong desire to achieve or to become sth 渴望；有志于

omni-channel /ˌɒmnɪˈtʃænəl/ *n.* used to refer to a way of selling products that is the same and equally good for the customer whether they are buying from a computer, a mobile phone app, etc, or in a physical shop 全渠道；全方位

embed ... into to fix sth firmly into a substance or solid object 把……牢牢地嵌入（或插入、埋入）

seek to	try to do sth 力图；设法
in addition	used when you want to mention another person or thing after sth else 除……以外（还）
depend on	according to 视乎；决定于
combine ... with	to put two or more different things, features, of qualities together 结合，融合
attach importance to	to believe that sth is important or worth thinking about 认为有重要性；重视
adhere to	to behave acoording to a paticular law, rule, set of instractions, exc.; to follow a particular set of beliefs or a fixed way of doing sth 坚持，遵守，遵循（法律，规章，指示，信念等）
at the forefront of	in or into an important or leading position in a particular group or activity 处于最前列；进入重要地位（或主要地位）

Reading Comprehension

Understanding the text

Choose the best answer to each of the following questions.

1. According to Paragraph 2, how has Urban Revivo positioned itself in the retail field?

   A. A pioneer of affordable yet trend-driven high-quality products.

   B. A pioneer of expensive and trend-driven high-quality products.

   C. A pioneer of affordable yet trend-driven medium-quality products.

   D. A pioneer of expensive and trend-driven medium-quality products.

2. Why was London chosen for Urban Revivo's European flagship?

   A. Because it is an international market with a large number of potential customers.

B. Because it is a landmark of international fashion and one of the most mature markets in the world.

C. Because it is the most mature market in the world and it has many famous designers.

D. Because it is an international city and will help Urban Revivo keep up with the changing fashion trends.

3. According to Leo Li, what will Urban Revivo do to further explore potential new markets?

A. To pay more attention to product development and improvement.

B. To continuously improve the quality of the store's operations.

C. To develop online channels and cooperate with European e-commerce platforms.

D. All of the above.

4. Which of the following is NOT done by Urban Revivo to enhance its fashion expression?

A. Gradual creation of the product.

B. Extremely flexible supply chain.

C. More styles and lower prices.

D. Efficient logistics.

5. According to Leo Li, what are the challenges confronting Urban Revivo?

A. Continuing to develop and expanding into a global market.

B. Keeping up with the changing fashion trends and attracting new customers.

C. Designing new styles and lowering prices.

D. Competing with other retail brands and showing its uniqueness.

Critical thinking

Work in pairs and discuss the following questions.

1. What's the impact of fast fashion on consumers and fashion industry?
2. What do you think are the future challenges for fast fashion?
3. According to the text, Urban Revivo has made attempts to challenge the traditional model of fast-fashion production. In your opinion, what other strategies can UR pursue to create an "international fashion empire" and rebuild the image of "made-in China"?

Research project

Slow fashion is fast fashion's sustainable alternative. Search for some information about it in terms of its concept, characteristics and potential influences. Write a research report to summarise your findings and present it in class.

Language Enhancement

Words in use

Fill in the blanks with the words given below. Change the form when necessary. Each word can be used only once.

position	represent	target	claim	diverse
challenge	omni-channel	era	element	rapidly

1. The ____________ was characterised by political and cultural turbulence.
2. It ____________ that the officers tortured a man to death in 1983 in a city police station.
3. The shop front occupies a very prominent____________ on the main street.
4. Finding a cure for cancer is one of the biggest ______________ facing medical researchers.
5. His law firm ____________ a dozen of the families involved in that disaster.
6. From the 1930s to the 1950s the incidence of walking to work declined ____________, but the use of buses and bicycles increased substantially.
7. The advertisement for the energy drink ____________ specifically at young people.
8. They've introduced all sorts of new ____________ to that programme in order to broaden its appeal.
9. We need to recognize that we live in a culturally ____________ society.
10. A new concept called ______________ retail was introduced to describe retail environment where all channels, both in online and offline are seamlessly integrated.

Expressions in use

Fill in blanks with the expressions given below. Change the form when necessary. Each expression can be used only once.

adhere to	embed into	attach importance to
at the cutting edge of	in addition	combine with
seek to	at the forefront of	

1. The dress________________ stylish lines ________________ an attractive floral print for a classically feminine look.
2. His research is ________________ new therapies for cancer.
3. The bank ________________ all rules regarding the opening and closing of accounts.
4. All employees receive paid holiday and sick leave. ________________, we offer a range of benefits for new parents.
5. His team is ________________ scientific research into vaccines.
6. I don't ________________ any ________________ these rumours.
7. The country's president is ________________ mend relations with the United States.
8. Microprocessors are ________________ products such as cars, fridges, traffic lights, and industrial equipment.

Sentence structure

I. Complete the following sentences by translating the Chinese into English, using "along with … " structure.

Model: They still see its home market of China as having great potential, ____（以及亚太地区和美国市场）.

→ They still see its home market of China as having great potential, along with the Asia-Pacific region and the US market.

1. She lost her job when the factory closed, ______________________________ (和成百上千的其他人一样).

2. In the 18th century art was seen, ______________________________ (以及音乐和诗歌), as something edifying.

3. Martin Luther King Junior also led the Montgomery bus boycott, ____________ ______________________________ (以及许多其他民权示威活动).

II. Rewrite the following sentences by using emphatic sentence structure "it is … that/who …".

Model: Play fashion is the core of our brand, and this essence has always been adhered to.

→ Play fashion is the core of our brand, and it is this essence that has always been adhered to.

1. This strategic ability of introducing new collections based on latest trends in a rapid manner enabled this fashion brand to beat other competitors.

____

____.

2. That tall man wrenched the handbag from the old woman last night.

____

____.

3. Stability destroys people's ambition and barricades people's steps.

____

____.

Extensive Reading

Louis Vuitton: The History of a Luxury Giant

1 As one of the longest standing luxury brands, Louis Vuitton is the standard of opulent fashion. With a signature monogram that is not only iconic but often replicated, it is easy to see why carrying a Louis Vuitton bag is a status symbol.

2 Strive not to be successful, but rather to be of value. Louis Vuitton Malletier, commonly known as Louis Vuitton is a popular French fashion house established by Louis Vuitton in Paris in 1854. The label's LV monogram appears on most of its products, ranging from luxury trunks and leather goods to ready-to-wear, shoes, watches, jewelry, accessories, sunglasses, and books. Louis Vuitton is one of the world's leading international fashion houses, selling its products through standalone boutiques, lease departments in high-end department stores, and through the e-commerce section of its website. For six consecutive years (2006–2012) Louis Vuitton has been named the world's most valuable luxury brand. Its 2012 valuation was 25.9 billion USD.

3 The success story of this brand goes back to a young man who left his hometown to Paris to try and make a new life for himself at the age of 16. At that time, the city was in the midst of industrialisation with current modes of transportation evolving quickly allowing for longer journeys. With this trend came the need for sturdy travel pieces. Vuitton was taken as an apprentice for a successful box maker and packer named Monsieur Maréchal. In 1854, years after he mastered the craft of making durable containers, Vuitton ventured out on his own to open a shop on Rue Neuve des Capucines. It was here that he began to establish himself as a luggage maker and started his own company, which was an immediate success.

In 1858, Vuitton designed the first Louis Vuitton steamer trunk, a design that is still used today. At the time trunks had rounded tops to allow for water to run off but this did not allow for convenient stowage. Vuitton introduced a flat, yet waterproof, trunk that was easily stackable. The first of his trunks were outfitted with a gray canvas referred to as Trianon—it wouldn't be until several decades later that the signature monogram would be introduced. With a burgeoning business, Vuitton moved his family and workplace to Asniere, where he employed twenty workers to craft his trunks. By 1900 he would have 100 employees, and in 1914 the company would more than double in size.

4 After years of success, Vuitton began to experiment with the design of his luggage by introducing a new striped canvas pattern (1876) and later the still well-known Damier print (1888). The hand-painted patterns were developed to prevent counterfeits. Even in the late 1800s, Louis Vuitton was enough of a status symbol to warrant counterfeiting. In 1886, his son George invented and patented an ingenious locking system that made it impossible to pick the lock of their trunks. This lock is still used today. 1892 would prove to be a time of mourning for the family as Louis Vuitton passed away at the age of 70. His son, Georges Vuitton, became the new head of the luxury house. Louis Vuitton's passing would prompt his son to once again change the print of their luggage, and in 1896, to honor his father, the signature LV monogram was introduced and patterned with LVs, quatrefoils, and flowers. Under his direction success followed and the iconic monogram rose to fame among elite clientele. With this rise in prominence, the brand caught the eye of one of the most established of fashion icons: Gabrielle Chanel.

5 Louis Vuitton has become the world's biggest and popular store and has conquered the world through its unique designs. In 1914, Vuitton opened the biggest luggage store in the world on the Champs-Elysées, becoming the favourite of millionaires and the elite. The company was capable of delivering every single category of luggage to its customers. Vuitton

monogram is one of the most luxurious and recognizable trademarks in the world. The company was also listed on the Paris Bourse in 1984. The company merged with many other popular brands like Veuve Clicquot and Loewe, etc. What was once a tony little Parisian luggage shop is now the multifaceted jewel in the crown that sits atop the head of Bernard Arnault, CEO of the fashion conglomerate LVMH, who, in 2003, likened the revenue-generating house to a "luxury Microsoft." Now is the time for the company to become one of the most luxurious brands all over the globe.

6 After a great success, many other brands started to market Louis Vuitton as their label brand. A wide range of men and women collection is also found under this brand, among which menswear is one of the important selling lines of Louis. The brand uses the finest fabric for its collection all over the world. Louis Vuitton is the world's 29th recognized brand with its net worth of around US $19 billion. Considered as the most powerful brand all over the globe, the best part of this brand is that the luggage is still made by hands only by experts. Famous for the creation of a limited-edition collection every season and collaborations with various renowned artists and designers, Louis has used a number of celebrities for its advertising and always created a range of unique collection.

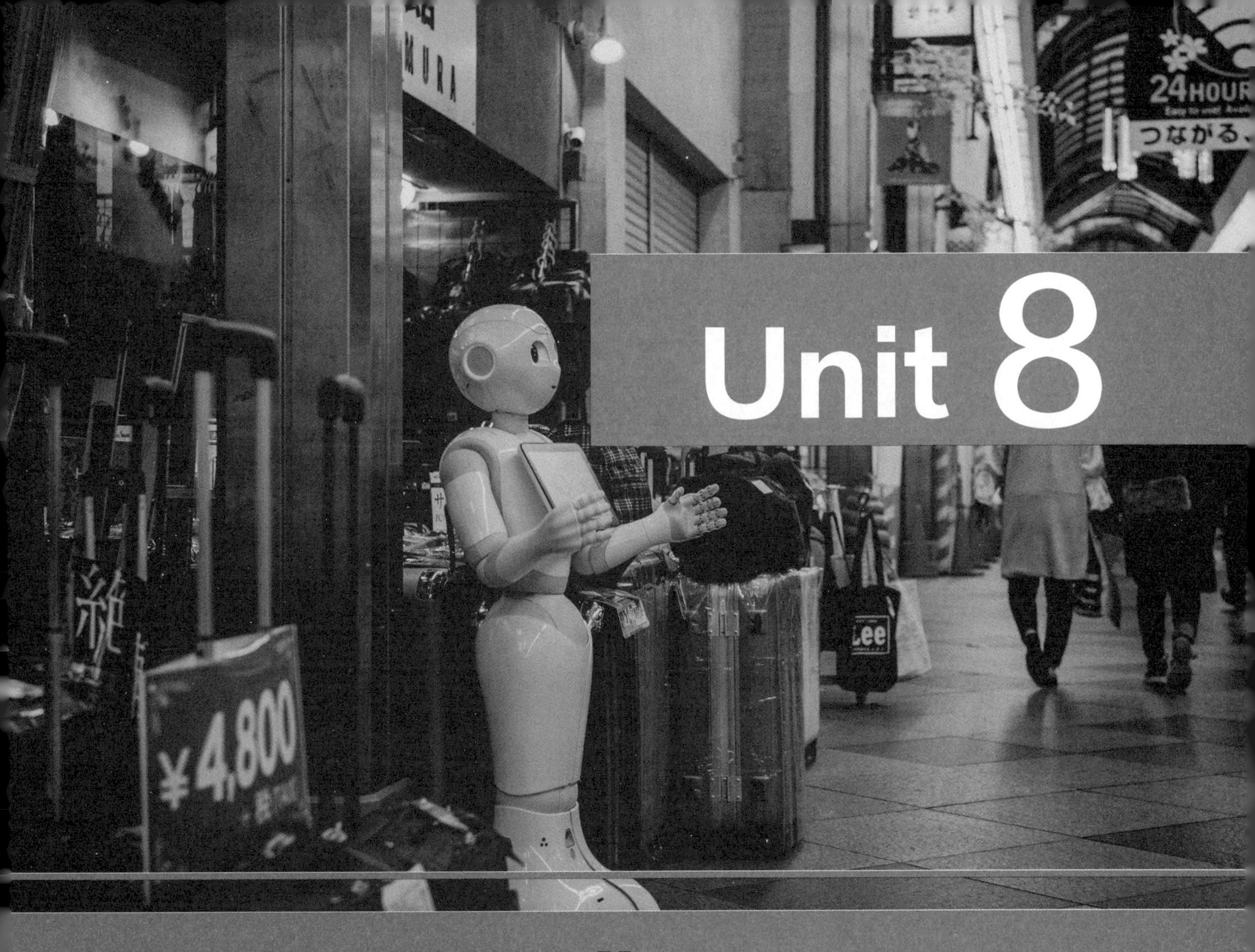

Unit 8

Artificial Intelligence and Fashion

Artificial Intelligence and fashion are two fields which have often misunderstood each other, but this is beginning to change as people recognize the creativity in both.

— *Anonymous*

Artificial intelligence (AI) is everywhere at the moment. Much of this is focused around data analytics—crunching business numbers at ferocious speed and as frequently as possible.

— *Malcolm Newbery*

Pre-Reading Activities

1. Watch the video about how AI technology facilitates Alibaba's business and fill in the blanks based on what you hear.

Alibaba has opened a new fashion AI concept store right next to Hong Kong Polytechnic universities fashion faculty, where (1) ________, engineers and fashionistas are working together to (2) ________ the future of fashion retail. Check into the concept store with a mobile Taobao ID (3) ________ and shoppers can opt in to a face (4) ________ for a more personalised experience. You can browse around like you would at any clothing store. The difference is that items you (5) ________ up from the RFID-enabled clothing rack will (6) ________ show up on the mirror. The smart mirror also shows personalised mix-and-match (7) ________ to complete the look. It also directs shoppers to wear other (8) ________ items can be found inside the store.

2. Work in pairs and discuss the following questions.

(1) Do you think that we will have more concept stores in the future? And why?

(2) What influences may a concept store exert on our society?

(3) Will its advantages outweigh its disadvantages? And why?

Bridging the Gap between Artificial Intelligence and Creativity in Fashion

1 The world's oldest known human drawing dates back 73,000 years to a

small stone in what is now South Africa, and the oldest known human painting dates back more than 40,000 years ago to Western Europe and Southeast Asia. Creativity in its purest form has been around nearly as long as modern humans: as defined by Margaret Boden in 1998, creativity is a **fundamental** activity of human information. It is characterised by two **branches**, generally speaking, "The ability to produce work that is both **novel** and **appropriate**." In **layman**'s terms, creativity is the act of creating that which is **original** and **adaptive**.

2 But when it comes to fashion, is creativity **inherently** human?

3 Nikoleta Kerinska, PhD in art sciences, stated, "Artificial intelligence can **simulate** creativity. But there is also the question of the artist's intention. The human artist **consciously** works toward a creative or **conceptual** goal. A computer program does not have this same consciousness."

4 The field of artificial intelligence as an **academic discipline** was first founded in 1956 at Dartmouth College in the United States. While advancements have been made in leaps and bounds since then, the definition of AI remains **debated**. *The English Oxford Living Dictionary* defines AI as such: the theory and development of computer systems able to perform tasks normally requiring human intelligence, such as visual **perception**, speech recognition, decision-making, and translation between languages. *Merriam Webster* gives a much more **concise** definition: the **capability** of a machine to **imitate** intelligent human behaviour. Yet there runs one common thread through all accepted definitions of AI today, and that is technology which imitates human intelligence to, in some cases, **outperform** the human brain. Methods such as deep learning and machine learning work to this end, and learn from repetition and input not unlike a human.

5 But technology that can imitate the thinking processes of a human can be an **unsettling** thought, especially when the question of what creativity is comes into the picture. Creativity is a sense of pride: for many, creativity

explains humankind's progress throughout history, and it is this critical thinking that sets humans apart from the rest. For those in the creative fields, there is often **suspicion** toward artificial intelligence, because its **connotations** of hard logic seem to take away from the **heart-on-sleeve** process of traditional creatives. Yet 100 years ago, the idea of using the Internet was **inconceivable**, and today it has become an extension of the creative process. The Internet hasn't replaced painting, or object design, or couture, has it? In the same way, artificial intelligence will be, and is already for some, a tool to take human creativity to new heights and to open doors we didn't know existed.

6 So if artificial intelligence isn't a distant future anymore, how is it being used within creative disciplines? Interestingly enough, AI is already being incorporated into many different artistic fields, including but not limited to painting, object design, and fashion design.

7 It's hard to imagine a painting without a **paintbrush** in a human hand, but AI comes with surprises. Last year in February, there was something special about an **exposition** presented at the Gallery Vossen in Paris. Unlike other paintings, these paintings were not painted by a human hand. Robbie Barrat, an artist and researcher within the field of AI, teamed up painter Ronan Barrot to produce a marriage of human creativity and AI technology. Using deep learning to feed over 500 paintings from Ronan Barrot into the machine, the pair presented their human-inspired, machine-painted art to the world. With some scepticism, the concept was **well-received**, a marriage of human **input** and tech **output**.

8 In object design, another pair of minds had the same idea of joining human input to produce an artificially-made result: enter the Chair Project, which was launched in 2018 between designers Philipp Schmitt and Steffen Weiss. The project was conceived with the goal of determining whether or not AI could indeed be as creative as a human designer, and the results were **fascinating**. They asked themselves: "Could a **bot** produce an object as **symbolically** rich as a chair, and with the aesthetic **proficiency** of an

Eames or Breuer design?" The pair created a **generative neural** network (GAN) and fed into it a **dataset** of over 500 chair designs from the 20th century. In the end, the designers refined the machine's creations down to 4 different chairs, all **unconventional**. So while AI cannot and should not replace designers, it might be a useful **collaborative** artistic tool, as evidenced by the Chair Project.

9 When it comes to **skepticism** toward creativity, AI is perhaps most **salient** within fashion, whether it be collection planning or design itself. The foundation has taken years to be built: for example, at Burberry's 2010 Fall/Winter show in London, the collection was streamed live using 3D technology to guests at private locations in New York, LA, Paris, Tokyo, and Dubai. At the time, this was **revolutionary**, and it seems the pandemic has brought back the **digital** fashion show full circle.

10 The gaming industry equally played its part in paving the way for fashion without even knowing it. Games such as *World of Warcraft* and *Fortnight* have been selling skins for over two decades, and few could have **predicted** this concept would one day be translated to the average consumer's real-life wardrobe. The 3D AI technology used for video games renders characters more lifelike with details such as natural bodily movements, facial expressions, and fabrics which flow in the wind or against the body. In turn, this technology is used more and more in fashion to analyse and reproduce the true nature of fabric and fit for consumers to create real-life or digital clothing. Additionally, the digital clothing industry is a **viable** solution to waste and pollution issues, and it fits in with the social media generations of today: it has the potential to represent 1% of the fashion market share at $25 billion.

11 Given the ways people have already adapted artificial intelligence to their creative **endeavors**, it becomes difficult to assert that AI and creativity are **mutually** exclusive. AI is human-created, after all, and this technology is a tool rather than a replacement: human creativity, like in thousand-year-old cave paintings, will always be the foundation of art. And with artificial intelligence, traditional creation is changing shape.

Notes

Artificial Intelligence (AI) AI refers to the ability of a digital computer or computer-controlled robot to perform tasks commonly associated with intelligent beings. The term is frequently applied to the project of developing systems endowed with the intellectual processes characteristic of humans.

Generative Neural Network (GAN) 生成对抗网络，是一种非监督式学习方法，通过让两个神经网络互相博弈的方式进行学习。

New words and phrases

fundamental /ˌfʌndəˈmentl/ *a.* ① of or forming the basis or foundation of sth; essential 基础的，构成基础的；根本的 ② most important; central or primary 十分重要的；主要的，首要的 ③ ~ **(to sth)** essential or necessary 根本的，必要的

n. [usually pl.] basic rule or principle; essential part 基本规则 / 原则；核心部分

branch /brɑːntʃ/ *n.* [U] ① arm-like division of a tree, growing from the trunk or from a bough 树枝 ② subdivision of a family, a subject of knowledge, or a group of languages 家族的分支；知识的分科；语言的分系

novel /ˈnɔvəl/ *a.* new and strange; of a kind not known before 新奇的；新颖的

n. book-length story in prose about either imaginary or historical characters 长篇小说

appropriate /əˈprəupriət/ *a.* ~ **(for/to sth)** suitable; right and proper 适当的；合适的，正当的

layman /ˈleimən/ *n.* [C] a person who does not have expert knowledge of a particular subject 外行，门外汉

original /əˈridʒənəl/ *a.* able to produce new ideas; creative 有创见的；创造性的

adaptive /əd'æptɪv/ *a.* having the ability or tendency to adapt to different situations 有适应不同情况能力的

inherently /ɪnˈhɪərəntli/ *ad.* having a quality that is inherent in sth or is a natural part of it and cannot be separated from it 内在地，固有地

simulate /ˈsɪmjʊleɪt/ *v.* reproduce (certain conditions) by means of a model, etc., eg. for study or training purposes (用模型等)模拟某环境(如用于研究或训练)

consciously /ˈkɒnʃəsli/ *ad.* (of actions, feelings, etc.) realized by oneself; intentional (指行为、感情等)自觉地；蓄意地

conceptual /kənˈseptʃuəl/ *a.* consisting of, or relating to concepts or conception 概念的；与之有关的或组成概念的

academic /ˌækəˈdemɪk/ *a.* ① of (teaching or learning in) schools, colleges, etc. 学校的；学院的 ② scholarly; not technical or practical 学者式的；非技术的或实用的

discipline /ˈdɪsɪplɪn/ *n.* ① [C] branch of knowledge; subject of instruction 学科；教学科目 ② [U] result of such training; ordered behaviour 纪律

debate /dɪˈbeɪt/ *n.* [C, U] formal argument or discussion of a question, eg. at a public meeting or in parliament, with two or more opposing speakers, and often ending in a vote 正式的辩论，讨论(如在公众集会或议会中，常以表决结束)

perception /pəˈsepʃən/ — *n.* ① [U] ability to see, hear or understand 感知能力；认识能力② [C] ~ **(that ...)** way of seeing or understanding sth 看法，理解

concise /kənˈsaɪs/ — *a.* (of speech or writing) giving a lot of information in few words; brief (指语言或文字) 用少数词语传达大量信息的；简明的

capability /ˌkeɪpəˈbɪlɪti/ — *n.* [U] ~ **(to do sth/of doing sth)** | ~ **(for sth)** quality of being able to do sth; ability 能做某事的素质；能力

imitate /ˈɪmɪteɪt/ — *v.* copy the behaviour of (sb/sth); take or follow as an example 学某人 / 物的样；仿效

outperform /ˌautpəˈfɔːm/ — *v.* to achieve better results than sb/sth 优过，胜于

unsettling /ʌnˈsetlɪŋ/ — *a.* making sb feel upset, nervous or worried 让人沮丧的，紧张的，担心的

suspicion /səˈspɪʃn/ — *n.* [U] suspecting or being suspected 怀疑；涉嫌

connotation /ˌkɔnəˈteɪʃn/ — *n.* [U] an idea suggested by a word in addition to its main meaning 主要含义

heart-on-sleeve /hɑːt ɔn sliːv/ — *a.* show feelings and emotions rather than keeping them hidden 直白的

inconceivable /ˌɪnkənˈsiːvəbəl/ — *a.* sth that cannot be imagined; not conceivable 不可想像的；匪夷所思的

paintbrush /ˈpeɪntbrʌʃ/ — *n.* a brush that is used for painting 画笔

exposition /ˌekspəˈzɪʃn/ — *n.* ① [U] explaining or making clear by giving details 解释；说明 ② [C] explanation of a theory, plan, etc. 说明；解说

well-received /ˌwelˈrɪˈsiːvd/ — *a.* getting a good reaction from people 受到好评

input /ˈɪnput/ — *n.* [U] action of putting sth in 放入；投入；输入

output /ˈautput/ — *n.* [U] the amount of sth that a person or thing produces

fascinating /ˈfæsɪneɪtɪŋ/ — *a.* extremely interesting and attractive 极其有趣，吸引人

bot /bɔt/ — *n.* [C] abbreviation for robot 机器人的缩写

symbolically /sɪmˈbɒlɪkəli/ — *ad.* sth that is used to represent a quality or idea 象征地；象征意义地

proficiency /prəˈfɪʃənsi/ — *n.* [C] a good standard of ability and skill 精通，熟练

generative /ˈdʒenərətɪv/ — *a.* able to produce; productive 能生产的；有生产力的

neural /ˈnjuərəl/ — *a.* of the nerves 神经的

dataset /ˈdeɪtəˌset/ — *n.* [C] a collection of data 数据集

unconventional /ˌʌnkənˈvenʃnəl/ — *a.* very different from the way people usually behave, think, dress etc. 不依惯例的，非传统的，非常规的

collaborative /kəˈlæbərətɪv/ — *a.* [only before noun] (*formal*) involving, or done by, several people or groups of people working together 合作的

skepticism /ˈskeptɪsɪzəm/ — *n.* [U] an attitude of doubting that particular claims or statements are true or that sth will happen 怀疑论

salient /ˈseɪliənt/ — *a.* most noticeable or important; main 显著的，最重要的；主要的

revolutionary /ˌrevəˈluːʃənəri/ *a.* completely new and different, especially in a way that leads to great improvement 革命的，革新的

digital /ˈdɪdʒɪtl/ *a.* describes information, music, an image, etc. that is recorded or broadcast using computer technology 数字的

predict /prɪˈdɪkt/ *v.* to say that sth will happen in the future 预言

viable /ˈvaɪəbəl/ *a.* sound and workable; feasible 切实可行的；可实施的

endeavor /ɪnˈdevə/ *n.* [C, U] (*formal*) an attempt to do sth new or difficult 努力，尽力
v. (*formal*) try very hard ~ **to do sth** 努力做某事

mutually /ˈmjuːtʃuəli/ *ad.* felt or done equally by two or more people 相互地

be characterised by have sth as a typical quality or feature 以……为特点，具有……特征

when it comes to identify the specific topic that is being talked about 谈到，说到

in leaps and bounds sb or sth improving or increasing quickly and greatly 进展或进步很快

a sense of pride a feeling of satisfaction 自豪感；荣誉感

set apart from ①to move sth so it is away from sth else 把……和……分隔开 ② to make sth stand out when compared to sth else 让……脱颖而出

be incorporated into to include sth as part of sth larger 合并

fit in to belong to a group, plan, or situation 适应，适合

be the foundation of	to provide the conditions that will make it possible for sth to be successful 成为……的基础

Reading Comprehension

Understanding the text

Answer the following questions.

1. What is creativity in layman's terms according to the author?
2. What is the difference between a human artist and a computer program?
3. What is the common thread concerned with the definitions of AI?
4. Why do people often feel suspected toward the creativity of artificial intelligence?
5. What is the goal of the Chair Project?
6. How do you understand the first sentence in Paragraph 9?
7. In the last paragraph, why does the author say "it becomes difficult to assert that AI and creativity are mutually exclusive"?
8. Do you think AI will stifle human's creativity? And why?

Critical thinking

Work in pairs and discuss the following questions.

1. Apart from selling services provided by artificial intelligence mentioned in the text, what other benefits can AI bring to us in clothing and fashion industry?
2. Do you have any shopping experience that is shaped by the aid of AI technology? How do you feel about it? And in what aspects can it be improved in the future?

Research project

Suppose you are given a chance to start a clothing company, how do you plan to apply technology to facilitate your business? Write a detailed proposal and present it in class.

Language Enhancement

Words in use

Fill in the blanks with the words given below. Change the form when it is necessary. Each word can be used only once.

adaptive	perception	simulate	imitate	layman
proficiency	appropriate	suspicion	discipline	unsettling

1. The book was written in a style ________ to the age of the children.
2. To the ________, all these plants look pretty similar.
3. Societies need to develop highly ________ behavioural rules for survival.
4. Some driving teachers use computers to ________ different road conditions for learners to practice on.
5. Lack of ________ at home meant that many pupils found it difficult to settle in to the ordered environment of the school.
6. He is interested in how our ________ of death affect the way we live.
7. He was a splendid mimic and loved to ________ Winston Churchill.
8. The country's economic crisis had an ________ effect on world markets.
9. Evidence of basic ________ in English is part of the admission requirement.
10. He was arrested on ________ of having stolen the money.

Banked cloze

Fill in the blanks by selecting suitable words from the word bank. You may not use any of the words more than once.

A. honors	F. cutting edge	K. institutions
B. cast	G. patented	L. recognized
C. rigorous	H. doubted	M. judged
D. innovative	I. nominated	N. traditional
E. commercial	J. notorious	O. opportunities

"Smart" clothes that can cool and heat and a new superfilter that will make drinking water safer are two 1. ________ results of space technologies developed at NASA's Johnson Space Center recently 2. ________ by the U.S. Space Foundation.

The 3. ________ technologies are among those named to the Space Foundation's Space Technology Hall of Fame for 2005. Established in 1988 jointly by NASA and the Space Foundation, the Space Technology Hall of Fame 4. ________ technologies developed for space which have found commercial applications. Each year, technologies are 5. ________ and go through a 6. ________ selection process before final selection and induction.

Both JSC technologies honored this year were funded by NASA's Small Business Innovation Research and the Small Business Technology Transfer programs. Those programs provide 7. ________ for small companies and research 8. ________ to partner with NASA for research and development.

"These are 9. ________ technologies, things that seem like science fiction but are now in commercial use, improving lives on Earth," said Dr. Kumar Krishen, chief technologist and manager of the small business programs at JSC. "The development of 10. ________ technologies like this is needed to enable the vision for space exploration to return to the moon and travel beyond."

Expressions in use

Fill in the blanks with the expressions given below. Change the form when necessary. Each expression can be used only once.

in leaps and bounds	incorporate into	a sense of pride
when it comes to	set apart from	be characterised by
be the foundation of	fit in	

1. This weekend will ________ increasing heat and humidity.
2. ________ playing chess, he's the best I know.
3. Her Spanish has improved ________ this year.

4. Yet the project gave the majority of participants ________ in their work, connectedness, and, above all, enjoyment.
5. What ________ her ________ the other candidates for the job was that she had a lot of original ideas.
6. Suggestions from the survey have been ________ the final design.
7. Working in a research lab really ________ with my shy personality.
8. This theory will also ________ the modern revolution in our understanding of the deepest parts of the earth.

Translation

Ⅰ. Translate the following paragraph into Chinese.

Further technological refinement and AI development will allow fashion labels to become more active in merchandising products through online avatars, such as video games and digital communities. With these channels offering brands untapped access to legions of loyal followers, high-quality 3D visualisations can open up new horizons for fashion label marketing departments. It seems only a matter of time before 3D visualisation and virtual reality technology become the standard over physical products and samples when presenting to consumers in B2C environments. With key fashion technology players moving forward faster than the industry can keep pace, the road ahead seems both uncertain and exciting.

Ⅱ. Translate the following paragraph into English.

VR(AR/AI)技术在时尚零售产业中的使用虽然刚刚起步，但拥有巨大的潜力。习惯了传统购物方式的人们更倾向看得见摸得着的实体服装，这种虚拟服装可能给他们带来的只是一种怪异的体验。但是，越来越多的时装品牌对使用虚拟化身（avatars）作为时尚影响者和虚拟服装设计的想法持开放态度，他们通过在AI和混合现实中创新作品来回应他们的追随者和购物者。

Paragraph Writing

How to Properly Cite Others' Work?

Writing is a process of standing on the shoulder of giants and reflect on your own life. So, when you write, you will be inevitably encounter with other people's work especially if you are dealing with academic writing. How to cite other authors' work is a big issue in writing, for if not handled properly, you will be accused of "plagiarism." According to *Cambridge Dictionary*, plagiarism means "the process or practice of using another person's idea or work and pretending that it is your own." Although many students are aware of the ethical issues behind plagiarism, many students plagiarise unconsciously.

Apart from "apparent" plagiarism, like "Copying large pieces of text from a source without citing that source," you should pay special attention to the following cases:

1. Mentioning an author or source within your paper without including a full citation in your bibliography.
2. Citing a source with inaccurate information, making it impossible to find that source.
3. Using a direct quote from a source, citing that source but failing to put quotation marks around the copied text.
4. Using translation tools to translate the source language into your target language or vice versa without mentioning the source.

So when you write, you should be careful about the citing issue. At present, most countries follow the APA style. You can find the reference examples and paper format on the website: https://apastyle.apa.org.

As mentioned above, direct citing is one scenario. In English writing, you may also cite indirectly. Paraphrasing is one of the most common method for indirect citing. As a matter of fact, since paraphrase gives more freedom to synthesise, compare and contrast with relevant information, it is highly recommended by teachers and experts when you do paperwork. The basic idea of paraphrasing is to "restate a text, passage, or work by giving the meaning in another form." Here is an example of paraphrasing.

The original sentence:

"No historian should begin research with someone else's notes. Taking notes is the first (and perhaps most important step) in developing our own interpretation of a subject. It forces us to decide (again and again) what is interesting and important" (Reuben, 2005, p. 413).

Sentence after paraphrase:

Reuben (2005) states that in order to develop an original analysis of their topic, historians must commence research with their own "notes" rather than relying on a secondary analysis of another's.

Try to paraphrase the following sentences.

1. The student requested that the professor excuses her absence, but the professor refused.
2. There will be a music concert next to Vienna coffee shop. Would you like to go?
3. International Center is hosting English Conversation classes. They help non-native speakers of English practice their English speaking skills.
4. The office of International Students and Scholars (ISS) at Purdue University is located in Schleman Hall.
5. The car that was pulled over by the police officer yesterday just had an accident. That driver is not careful.

The Fashion Industry is Getting More Intelligent with AI

1 As long as humans have started to wear clothes, we'd have the desire to express our own individuality, and one way to achieve that is through

fashion. The fashion industry is one of the biggest in the world, **estimated** at about 3 trillion dollars as of 2018, representing 2 percent of global GDP. Much of **brick-and-mortar** traditional retail as well as online **e-commerce** is dedicated to the sale of clothing and fashion items. So much so that Amazon acquired shoe **retailer** Zappos for $1 Billion in 2010, and major retailers such as Walmart, Target, Amazon, and others have themselves entered into the fashion retail business through their own brands and brand **partnerships**. Despite the established nature of the fashion industry, AI is fundamentally **transforming** the industry from the way that fashion companies manufacture their products to the way they are marketed and sold. AI technologies are transforming the fashion industry in every element of its value chain such as designing, manufacturing, **logistics**, marketing and sales.

2 The fashion industry is just as much about creating demand and brand awareness as it is about the manufacturing of fashion products. Clothing and **apparel** brands are constantly looking for new ways to get their goods in front of buyers and create awareness and demand in the market. Increasingly, fashion brands are using AI and machine learning to **maximise** users' shopping experience, improve the efficiency of sales systems through intelligent **automation**, and **enhance** the sales processes using predictive analytics and guided sales processes.

3 Fashion brands are also starting to **leverage** conversational **assistants** through **chatbots** and voice assistant devices such as Amazon Alexa, Apple Siri, Google Home, and Microsoft Cortana. Using conversational **interfaces**, fashion brands can gather data by asking customers questions, understanding customer desires and trends, diving deeper into their purchase patterns, and suggesting related and **add-on** items. For example, when a customer needs new shoes or a dress, instead of interacting with a website or mobile app, they can simply have a conversation with an intelligent conversational **agent**. Through back and forth dialog, the customer can find the **optimal** fashion product or **accessory** item. This

interaction provides greater satisfaction for the customer and much more valuable information for the fashion brand.

4 In addition to conversational systems, AI is making its way into ecommerce and mobile apps. Customers are now able to take pictures of clothing they like or styles they want to imitate, and smart image recognition systems can match the photos to real life items available for sale. Additionally, AI-enabled shopping apps allow customers to take **screenshots** of clothes they see online, identify shoppable apparels and accessories in that photo, and then find the same outfit and shop for similar styles.

5 In the documentary *Minimalism*, they share that there can be up to 52 seasons for clothing. Given the constant changes in fashion and design, retailers need to consistently keep up with the most current trends and predict consumer **preferences** for next season. Traditionally, retailers base their estimate of current year's sales on data from the **prior** year. But this is not always **accurate** because sales can be influenced by many factors that are hard to predict, such as changing trends. AI-based **approaches** for demand projection, however, can reduce **forecasting** error by as much as 50 percent.

6 Once the clothes are designed, AI technologies can also play a role in textile manufacturing. Fashion manufacturers are innovating the use of AI to help improve efficiency of manufacturing processes and **augment** human textile employees. AI systems are being used to **spot defects** in fabric and ensure that the colours of the finished textile match with the originally designed colours. AI technologies such as computer vision technologies are allowing quality **assurance** processes to be more streamlined.

7 Whereas it used to be that only ecommerce giants such as Amazon and Walmart used machine learning **algorithms** to figure out sales trends, now small retailers are also leveraging machine learning to understand this **dynamic** fashion market, which may provide them a better chance

to succeed. Intelligent, AI-enabled systems can also help provide greater intelligence for fashion brands by identifying patterns and predictive analytics that can provide insight into fashion trends, purchase patterns, and inventory-related guidance. One company at the **forefront** in innovation with AI applied to fashion is Stitch Fix, an online personal styling service. The company is using machine learning algorithms to provide better customer experiences for customers and make their supply chain more efficient.

8 Machine learning technologies are also being applied to expediting logistics and making the supply chain more efficient. AI is being used to manage and **optimise** supply chains as well as reduce shipping costs and **transit** time. Machine learning algorithms are being used to make more accurate predictions of inventory demand and therefore reduce **wastage** or **eliminate** last minute purchases to meet unexpected **spikes** in demand.

9 Computer vision enabled by machine learning is also being used to help spot fashion **fakes** and **counterfeit** products. Previously, spotting fakes required the trained eye of specialised customs or other **enforcement** officers. Now, AI systems can keep a consistent watchful eye on counterfeit products that look increasingly similar to the real ones. In this area, AI technologies are being applied by customs and border enforcement to help spot the **validity** of **high-end** products which are frequently counterfeited such as **purses** and sunglasses.

10 We are now seeing that AI technologies can add value in every part of the fashion industry, from the design process and manufacturing processes to sales and marketing of finished goods. The future of fashion is intelligent for sure.

Notes

Machine Learning Machine learning is a branch of artificial intelligence (AI) focused on building applications that learn from data and improve their accuracy over time without being programmed to do so.

Zappos 西班牙语，意为“鞋”，是一家美国在线鞋类和服装零售商

Walmart 美国跨国零售企业沃尔玛

Target 美国塔吉特百货公司

Amazon Alexa 亚马逊出品的智能语音助手

Apple Siri 苹果公司出品的智能语音助手

Google Home 由谷歌公司开发的智能家居盒子

Microsoft Cortana 微软出品的智能语音助手

Stitch Fix 美国服装电商，以利用数据科学为用户提供定制服务为企业营销模式

New words and phrases

estimate /ˈestɪmɪt/	*v.* judge or calculate the approximate size, cost, value, etc. of sth 估计；估价
brick-and-mortar /ˈbrɪk ənd ˈmɔːtə/	*a.* existing as a physical building, especially a shop, rather than doing business only on the internet 实体的
e-commerce /ˈiː kɔməːs/	*n.* [U] buying and selling goods on the Internet 电子商务
retailer /ˈriːteɪlə/	*n.* [U] the activity of selling goods to the public, usually in small quantities 零售
partnership /ˈpɑːtnəʃɪp/	*n.* [C, U] ~ **(with sb)** a relationship between two people, organisations, etc.; the state of having this relationship 合作伙伴关系
transform /trænsˈfɔːm/	*v.* ~ **sth/sb (from sth) (into sth)** completely change the appearance or character of sth/sb 完全改变某事物（或某人）的外观或特性

logistics /ləˈdʒɪstɪks/ *n.* [U] organisation of supplies and services, etc. for any complex operation 后勤

apparel /əˈpærəl/ *n.* [U] clothing; dress 衣服；服装

maximise /ˈmæksɪmaɪz/ *v.* increase (sth) as much as possible 使（某事物）增至最大限度

automation /ˌɔːtəˈmeɪʃn/ *n.* [U] use of automatic equipment and machines to do work previously done by people 自动化（用自动设备和机器做以前需要人来做的工作）

enhance /ɪnˈhɑːns/ *v.* increase (the good qualities of sb/sth); make (sb/sth) look better 增强

leverage /ˈliːvərɪdʒ/ *n.* [U] (*formal*) the ability to influence what people do 影响力
v. to make money available to sb in order to invest or to buy sth such as a company 发挥资金的杠杆作用

assistant /əˈsɪstənt/ *n.* [C] a person who helps or supports sb, usually in their job 助手

chatbot /ˈtʃætbɔt/ *n.* [C] a computer program designed to have a conversation with a human being, especially over the Internet（尤指网上）聊天程序

interface /ˈɪntəfeɪs/ *n.* [U] surface common to two areas 界面，分界面

purchase /ˈpəːtʃɪs/ *n.* [U] (*formal*) (action of) buying sth 购买
v. **~ sth (with sth)** | **~ sth (for sb)** **buy sth** 购买某物

add-on /ˈædɔn/ | *n*. [C] a piece of equipment that can be connected to a computer to give it an extra use 附件

agent /ˈeɪdʒənt/ | *n*. [C] person who acts for, or manages the affairs of, other people in business, politics, etc. (商业、政治等方面的)代理人，经纪人

optimal /ˈɔptɪməl/ | *a*. best; most likely to bring success or advantage 最佳的，最优的

accessory /əkˈsesəri/ | *n*. [usually pl.] thing that is a useful or decorative extra but that is not essential; minor fitting or attachment 附属品；配件，附件

screenshot /ˈskri:nʃɔt/ | *n*. [C] an image of what is shown on a computer screen 截屏

minimalism /ˈmɪnɪməlɪzəm/ | *n*. [U] a style of art, design, music etc. that uses only a very few simple ideas or patterns 极简主义

preference /ˈprefərəns/ | *n*. [U] liking for sth (more than sth else) (与他物相较之)喜爱；偏爱

in ~ to sb/sth 更喜欢某人或某物

prior /ˈpraɪə/ | *n*. [U] coming before in time, order or importance 较早的；居先的；较重要的

~ to 在……之前

accurate /ˈækjʊrət/ | *a*. ① free from error 正确无误的 ② careful and exact 精确的；准确的

approach /əˈprəutʃ/ *v.* come near or nearer to (sb/sth) in space or time (在空间或时间上)接近，靠近
n. way of dealing with a person or thing 方法；手段

forecast /ˈfɔːkɑːst/ *v.* tell in advance (what is expected to happen); predict with the help of information 预报；预测

augment /ɔːgˈment/ *v.* (*formal*) make sth larger in number or size; increase 增多，增大；增加

spot /spɔt/ *n.* [C] small (usually round) mark different in colour, texture, etc. from the surface it is on 斑点(通常指圆的)
v. (not in the continuous tenses 不用于进行时态) pick out (one person or thing from many); catch sight of; recognize; discover (从许多人或事物中)找出，辨出，认出(某人或某事物)；瞥见；发觉

defect /dɪˈfekt/ *n.* [C] fault or lack that spoils a person or thing 缺点，不足之处，毛病，瑕疵

assurance /əˈʃuərəns/ *n.* [U] ① (also self-assurance) confident belief in one's own abilities and powers 自信；把握 ② statement expressing certainty about sth; promise 保证；担保

algorithm /ˈælgərɪðəm/ *n.* [C] (especially computing) a set of rules that must be followed when solving a particular problem 算法

dynamic /daɪˈnæmɪk/ *a.* ① of power or forces that produce movement 动力的 ② (of a person) energetic and forceful (指人)精力充沛的，有力的

forefront /ˈfɔːfrʌnt/ *n.* [sing.] **the ~ (of sth)** the most forward or important position or place 最前部；最重要之处

optimize /ˈɔptɪmaɪz/ *v.* (AmE also optimize) to make sth as good as it can be; to use sth in the best possible way 最优化

transit /ˈtrænsɪt/ *n.* [U] process of going or being taken or transported from one place to another 搬运；载运；运输

wastage /ˈweɪstɪdʒ/ *n.* [U] amount that is wasted 损耗量

eliminate /ɪˈlɪmɪneɪt/ *v.* ~ **sb/sth (from sth)** remove (esp. sb/sth that is not wanted or needed) 消除，清除，排除（尤指不必要或不需要的某人 / 物）

spike /spaɪk/ *n.* [C] hard thin pointed piece of metal, wood, etc; sharp point（金属、木质等的）尖状物，尖头

fake /feɪk/ *n.* [C] ① object that seems genuine but is not 赝品 ② person who tries to deceive by pretending to be what he is not 骗子；冒充者

counterfeit /ˈkauntəfɪt/ *a.* made or done so that it is very similar to another thing, in order to deceive; fake 伪造的，假冒的

v. copy or imitate (coins, handwriting, etc.) in order to deceive 伪造，仿造（钱币等）；模仿（笔迹等）

enforcement /ɪnˈfɔːsmənt/ *n.* [U] the process of making sure that sth happens, especially that people obey a law 执行，强制

validity /vəˈlɪdɪti/ — *n.* [U] the state of being legally or officially acceptable 合法，有效性

high-end /ˈhaɪənd/ — *a.* intended for people who what very good quality products and who do not mind how much they cost 高端的

purse /pɜːs/ — *n.* [C] a small bag in which women keep paper money, coins, cards 钱包

be dedicated to — to decide that sth will be used for (a special purpose); to use time, money, energy, attention, etc. for sth 致力于

interact with — to communicate with or react to 互动

make its way into — to start to make progress in a career or activity 有起色，有进步

available for — able to be obtained, taken, or used 能够获得的，能够触及的

keep up with — to do sth or move at an acceptable rate, or at the same rate as sb or sth else 跟上

match … with — to compare sth with sth else to see whether their parts correspond 匹配

provide insight into — give a chance to understand sth or learn more about it 有机会了解更多

keep an eye on — to pay continued close attention to (sth) for a particular purpose 持续关注

Reading Comprehension

Understanding the text

Choose the best answer to each of the following questions.

1. According to the passage, which of the following is not the way that AI and machine learning can be used?

   A. Maximise user's shopping experience.

   B. Improve the efficiency of sales systems.

   C. Enhance the sales processes.

   D. Provide convenience.

2. How do fashion brands gather information?

   A. By using robots.

   B. By using conversational interfaces.

   C. By using statistics.

   D. By using computer.

3. Why should retailers need to consistently keep up with the most current trends and predict consumer preferences for next season?

   A. Because constant change happens in fashion and design.

   B. Because they want their sales boom.

   C. Because they want more profit.

   D. Because they like the change.

4. In the last paragraph, why does the author believe "The future of fashion is intelligent for sure"?

   A. Because AI technology is penetrating into every corner of our life.

   B. Because we can't avoid the change.

   C. Because AI technology makes our life better.

   D. Because AI technology is developing at an astonishing speed.

5. What is the tone of the whole passage?

   A. Persuasive.

   B. Positive.

   C. Negative

   D. Neutral.

Research project

As artificial intelligence technology penetrates deeper into fashion industry, many people start to worry about its negative infldence. According to your understanding, what negative influence can it cast? And how to solve these problems? Discuss it with your panthers and try to give a presentation in class.

Language Enhancement

Word in use

Fill in the blanks with the words given below. Change the form when necessary. Each word can be used only once.

purchase	augment	estimate	validity	enhance
transform	assurance	approach	retailer	accurate

1. She ________ that the work would take three months.
2. High street ________ reported a marked increase in sales before Christmas.
3. The movie ________ her almost overnight from an unknown schoolgirl into a superstar.
4. This is an opportunity to ________ the reputation of the company.
5. If you are not satisfied with your ________ we will give you a full refund.
6. Scientists have found a more ________ way of dating cave paintings.
7. As you ________ the town the first building you see is the church.
8. While searching for a way to ________ the family income, she began making dolls.
9. Shocked by thc results of the elections, they now want to challenge the ________ of the vote.
10. He gave me his personal ________ that the vehicle was safe.

Expressions in use

Fill in the blanks with the expressions given below. Change the form when necessary. Each expression can be used only once.

available for	match with	make one's way into
interact with	provide insight into	be dedicated to
keep an eye on	keep up with	

1. The Boy Scouts organisation ____________ helping boys become moral and productive adults.
2. Dominique's teacher says that she ______________ well ____________ the other children.
3. Edward was just beginning to ____________ life.
4. The new release will become ____________ purchase in April of next year.
5. They pioneered the product, but now they have to ____________ the competition as regards innovation and price.
6. ____________ your scorecard ______________ mine to see whether there is any disagreement.
7. The new data ____________ the ways that stars like the Sun end their lives.
8. Can you ____________ the biscuits so they don't burn?

Sentence structure

I. Complete the following sentences by translating the Chinese into English, using "Whereas it used to be … " structure.

Model: ____

____

（过去只有像亚马逊和沃尔玛这样的电子商务巨头才使用机器学习算法来确定销售趋势）, now small retailers are also leveraging machine learning to understand this dynamic fashion market, which may provide them a better chance to succeed.

→ Whereas it used to be that only ecommerce giants such as

Amazon and Walmart used machine learning algorithms to figure out sales trends, now small retailers are also leveraging machine learning to understand this dynamic fashion market, which may provide them a better chance to succeed.

1. ____ (过去出门必须带现金), now you have multiple ways to pay your bills.
2. ____ (在我小时候这里是个农场), now it has been transformed into a well-equipped park.
3. Now this information is accessible to everyone, ______________________ (过去只有少数人才能获取这方面的信息)。

II. Rewrite the following sentences by using "so much so that ..."

Model: The sale of clothing and fashion items is quite lucrative, therefore, in 2010, Amazon paid $1 Billion for shoe retailer Zappos, and major retailers including Walmart, Target, Amazon have joined the fashion retail business through their own brands and brand partnerships.

→ So much so that Amazon acquired shoe retailer Zappos for $1 billion in 2010, and major retailers such as Walmart, Target, Amazon, and others have themselves entered into the fashion retail business through their own brands and brand partnerships.

1. Because of their concern over the television programs that are aired, many parents are deciding which programs they will allow their children to watch.

____

____

2. The twins are identical, thus their parents often get them mixed up.

____

____

3. When the patient returned from the trip, he was exhausted to the point of being unable to stand up.

____

____

How One Clothing Company Blends AI and Human Expertise

1 When we think about artificial intelligence, we often imagine ro-bots performing tasks on the warehouse or factory floor that were once exclusively the work of people. This conjures up the specter of lost jobs and upheaval for many workers. Yet, it can also seem a bit remote—something that will happen in "the future." But the future is a lot closer than many realise. It also looks more promising than many have predicted.

2 Stitch Fix provides a glimpse of how some businesses have already made use of AI-based machine learning to partner with employees for moreeffective solutions. A five-year-old online clothing retailer, its success in this area reveals how AI and people can work together, with each side focused on its unique strengths.

3 The company offers a subscription for clothing and styling service that delivers apparel to its customers' doors. But users of the service don't actually shop for clothes; in fact, Stitch Fix doesn't even have an online store. Instead, customers fill out style surveys, provide measurements, offer up Pinterest boards, and send in personal notes. Machine learning algorithms digest all of this eclectic and unstructured information. An interface communicates the algorithms' results along with more-nuanced data, such as the personal notes, to the company's fashion stylists, who then select five items from a variety of brands to send to the customer. Customers keep what they like and return anything that doesn't suit them.

4 Stitch Fix's approach illustrates three lessons about how to combine human expertise with AI systems. First, it's important to keep humans in the

business-process loop; machines can't do it alone. Second, companies can use machines to supercharge the productivity and effectiveness of workers in unprecedented ways. And third, various machine-learning techniques should be combined to effectively identify insights and foster innovation.

5 As research we've conducted across industry and academia shows, companies have an unprecedented opportunity to tap ongoing advances in AI and machine learning research to reinvent business processes. For instance, in analysing a five-year sample of almost 1,150 papers, we identified at least 12 techniques, visible in the chart below, that can be readily applied and combined with each other within a process. Stitch Fix is already applying several of these machine learning techniques in service delivery and R&D—and other companies can follow its lead.

6 Some extremely successful companies have made great use of recommendation engines to boost sales or improve customer satisfaction. When it comes to recommendations, is there room for improvement in the way Amazon and Netflix operate?

7 Stitch Fix, which lives and dies by the quality of its suggestions, has no choice but to do better. And it can't rely solely on machines to do this. The company collects as much information about a client as it can, in both structured and unstructured form. Structured data includes surveys with personal information such as body measurements and brand preferences. Unstructured data can be derived from social media accounts, such as Pinterest, or through online notes from people about why they are buying new clothes, such as a special occasion, a change of season, or because a certain new style caught their eye.

8 The automated recommendation system is at its best when dealing with structured data. But to make sense of unstructured data, people and their judgment are needed. Say a client wants a new pair of stylish jeans, an item that's notoriously tricky to fit right to a person's measurements. To start, the algorithm finds jeans (across a range of fabrics, styles, and even

sizes) that other clients with the same inseam decided to keep—a good indicator of fit.

9 Next, it's time to pick the actual pair of jeans to be shipped. This is up to the stylist, who takes into account a client's notes or the occasion for which the client is shopping. In addition, the stylist can include a personal note with the shipment, fostering a relationship, which Stitch Fix hopes will encourage even more useful feedback.This human-in-the-loop recommendation system uses multiple in-formation streams to help it improve. The algorithm absorbs feedback directly from the client—whether or not she or he (the company added men's options in late September) decided to keep an item of clothing. And the stylist improves and adjusts based on cues gleaned from client notes and with insights from previous interactions with the customer.

10 The company is testing natural language processing for reading and categorising notes from clients—whether it received positive or negative feedback, for instance, or whether a client wants a new outfit for a baby shower or for an important business meeting. Stylists help to identify and summarise textual information from clients and catch mistakes in categorisation. Because the algo-rithms are never far from human oversight, Stitch Fix can confidently test new machine learning technologies without worrying that the experimentation will disrupt the client experience.

11 The fashion industry is no stranger to fast cycles of learning. One of the great benefits Stitch Fix sees from collecting and analysing so much data is an ability to predict trends. For example, the company's engineers are developing machine learning classifiers to find trends by using the simple yes-or-no decision that a client makes when they buy an item or send it back. From this seemingly simple data, the team has been able to uncover which trends change with the seasons and which fashions are going out of style.

12 We're only at the beginning of the era of artificial intelligence. Some

upheaval is to be expected. But we are starting to see how AI can change industries, improve productivity, and even benefit a new generation of employees.

Glossary

A			
		amass	Unit 4B
abhor	Unit 4B	ambitious	Unit 4A
abrasion	Unit 2A	anathema	Unit 4B
absorbent	Unit 2A	ancestor	Unit 1B
abstraction	Unit 4A	annually	Unit 7B
academic	Unit 8A	anthropologist	Unit 1B
accessible	Unit 5B	anticipate	Unit 4A
accessorize	Unit 5A	anticipation	Unit 6B
accessory	Unit 8B	apparel	Unit 2B
accurate	Unit 8B	application	Unit 2A
acid	Unit 2B	appreciate	Unit 6A
acknowledgement	Unit 1A	appreciation	Unit 4A
acoustic	Unit 2B	approach	Unit 8B
acquire	Unit 7B	appropriate	Unit 8A
adaptability	Unit 5B	archaeological	Unit 1B
adaptive	Unit 8A	architectural	Unit 6A
add-on	Unit 8B	arise	Unit 3B
adore	Unit 6B	articulation	Unit 6A
aesthetic	Unit 1B	artistry	Unit 5A
affiliated	Unit 7A	aspire	Unit 7B
affluent	Unit 3B	aspiring	Unit 6A
affordable	Unit 7B	assistant	Unit 8B
aftermath	Unit 5A	assurance	Unit 8B
agent	Unit 8B	atelier	Unit 4B
algorithm	Unit 8B	attempt	Unit 3A
alter	Unit 6B	attire	Unit 3A
alternate	Unit 4A	augment	Unit 8B
alternative	Unit 1B	autodidact	Unit 4B

automation	Unit 8B
avant-garde	Unit 4A

B

backlog	Unit 7A
baroque	Unit 3A
battlefield	Unit 7A
bead	Unit 3A
beneficial	Unit 2B
blazer	Unit 4B
bleach	Unit 1A
blogger	Unit 6B
bobbin	Unit 3B
bobbin lace	Unit 3B
bohemian	Unit 4B
bold	Unit 4A
bolster	Unit 7A
boom	Unit 1A
boon	Unit 5A
boost	Unit 4A
bot	Unit 8A
botanical	Unit 4A
boutique	Unit 5A
branch	Unit 8A
breakthrough	Unit 2B
breeches	Unit 5B
brick-and-mortar	Unit 8B
brilliant	Unit 4B
burgeoning	Unit 7A
buzz	Unit 6B

C

campaign	Unit 7A
candidly	Unit 4B
canvas	Unit 3A
capability	Unit 8A
capital	Unit 1A
capitalization	Unit 7A
capture	Unit 7B
caricaturist	Unit 4B
carve	Unit 7A
catapult	Unit 2B
category	Unit 7A
celebrity	Unit 5A
cellulose	Unit 2A
cement	Unit 5A
challenge	Unit 7B
charismatic	Unit 4B
chatbot	Unit 8B
chemical	Unit 2A
chintz	Unit 4A
civilization	Unit 1B
claim	Unit 7B
coarseness	Unit 2B
cocoon	Unit 2A
coincide	Unit 4A
collaborate	Unit 4A
collaborative	Unit 8A
collar	Unit 3B
coloration	Unit 1B
combination	Unit 5B
commercialise	Unit 1A
commission	Unit 4A
committed	Unit 4A
competitiveness	Unit 7B
compile	Unit 6B

complement	Unit 4A
composite	Unit 2A
composition	Unit 5B
compromise	Unit 2B
conceive	Unit 4A
concept	Unit 7A
conception	Unit 6A
conceptual	Unit 8A
concise	Unit 8A
confection	Unit 2A
confirm	Unit 4B
confront	Unit 7B
connotation	Unit 8A
consciously	Unit 8A
consequently	Unit 4A
conservatism	Unit 5B
consistent	Unit 3B
consolidate	Unit 1A
consortium	Unit 7A
conspicuously	Unit 5A
conspiratorial	Unit 4B
constantly	Unit 7B
consumption	Unit 7A
contemporary	Unit 4A
continously	Unit 7B
contribute	Unit 1B
convention	Unit 6A
convey	Unit 6A
cordage	Unit 2B
core	Unit 7B
corset	Unit 5B
costume	Unit 3B
couch	Unit 3B
counterfeit	Unit 8B
counterpart	Unit 7A
court	Unit 5A
couture	Unit 4B
craft	Unit 3B
cravat	Unit 5B
credential	Unit 1A
crinoline	Unit 5A
crochet	Unit 3A
crown	Unit 5A
crucial	Unit 1A
cuff	Unit 3B
cultivate	Unit 4B
curve	Unit 1B
cutaway	Unit 5B
cutting-edge	Unit 7B

D

dataset	Unit 8A
debate	Unit 8A
decade	Unit 3A
decent	Unit 7A
declare	Unit 5B
declining	Unit 5A
décor	Unit 2B
defect	Unit 8B
deferred	Unit 4A
definite	Unit 3B
deforestation	Unit 2B
degum	Unit 2B
democratise	Unit 5A
demonstrate	Unit 3B

density	Unit 2B
dependency	Unit 1A
depict	Unit 3A
deploy	Unit 6B
derive	Unit 5B
desirable	Unit 2B
despise	Unit 5B
despite	Unit 7B
detailing	Unit 6A
determine	Unit 1B
devilry	Unit 4B
diagonal	Unit 4A
digital	Unit 6B
digitally	Unit 7B
discipline	Unit 8A
discourse	Unit 6A
disguise	Unit 6A
displace	Unit 3B
display	Unit 3B
disseminate	Unit 3B
dissolve	Unit 2A
distinctive	Unit 5A
distort	Unit 2B
distribution	Unit 7A
diverse	Unit 7B
diversify	Unit 7A
domestic	Unit 7A
dominant	Unit 7A
dominate	Unit 1A
don	Unit 6B
downside	Unit 7A
dramatically	Unit 6B
drape	Unit 1B
draper	Unit 1A
drapery	Unit 3A
drastically	Unit 1B
drool	Unit 6A
durability	Unit 2B
durable	Unit 2A
dwindle	Unit 3A
dyer	Unit 1A
dynamic	Unit 8B

E

eccentric	Unit 5B
ecologically	Unit 2B
e-commerce	Unit 8B
economical	Unit 3A
edge	Unit 3B
edgy	Unit 6A
efficient	Unit 7B
elaborate	Unit 1A
elaborately	Unit 5A
elastic	Unit 2A
elasticity	Unit 2A
elegant	Unit 5B
element	Unit 7B
elevate	Unit 4A
eliminate	Unit 8B
eloquence	Unit 4B
embed	Unit 1A
embellishment	Unit 5A
embroider	Unit 3A
embroidered	Unit 3A
emerge	Unit 1B

empire Unit 7B
employ Unit 3A
enable Unit 7B
enamel Unit 3A
encapsulate Unit 4A
encyclopedic Unit 4B
endeavour Unit 6A
endorse Unit 6B
endure Unit 1A
enemy Unit 5B
enforcement Unit 8B
enhance Unit 8B
enormous Unit 1B
enterprising Unit 3B
enthusiasm Unit 3A
enthusiast Unit 6B
entrepreneur Unit 3B
entrust Unit 4B
entwine Unit 1A
enzymatic Unit 2B
equestrian Unit 5B
era Unit 7B
escapism Unit 6A
eschew Unit 5A
essence Unit 7B
essentially Unit 6A
establish Unit 5B
estimate Unit 8B
ethnic Unit 5B
etiquette Unit 5B
everlasting Unit 6B
evolution Unit 1B
evolve Unit 3B
exaggerate Unit 6A
exaggeration Unit 5B
exceptionally Unit 2B
exist Unit 3B
exorcise Unit 4B
expand Unit 7B
expansion Unit 7A
expedite Unit 1A
expendable Unit 7A
expose Unit 1B
exposition Unit 8A
extant Unit 1A
extract Unit 2A

F

fabled Unit 4B
fabric Unit 1B
fabrication Unit 4B
fake Unit 8B
fanatical Unit 6B
fanaticism Unit 6B
fanbase Unit 6B
fantasy Unit 6A
far-reaching Unit 4A
fascinating Unit 8A
fascinating Unit 6A
fashion Unit 1B
fashionable Unit 6B
fashionista Unit 6A
feed Unit 6B
felt Unit 1B
fibrous Unit 2A

fierce Unit 4B
figurine Unit 1B
filament Unit 2A
flagship Unit 7B
flamboyance Unit 5B
flax Unit 1B
flexibility Unit 7B
flexible Unit 2A
flog Unit 7A
floral Unit 5A
flounce Unit 5A
flourish Unit 3A
fluctuation Unit 4B
fluidly Unit 4A
foothold Unit 7A
foray Unit 7B
forecast Unit 8B
forefront Unit 8B
forthcoming Unit 6B
fossilise Unit 1B
fossilised Unit 3A
foster Unit 3A
frame Unit 6A
freehand Unit?3A
freelance Unit 4A
freshness Unit 7B
frock Unit 5B
frock coat Unit 5B
fuller Unit 1A
fundamental Unit 8A
fundamentally Unit 6A
furnishing Unit 4A
furrier Unit 4B

G

garment Unit 5B
garner Unit 7A
generate Unit 7A
generative Unit 8A
genetic Unit 1B
geographic Unit 6B
geometric Unit 3B
glimpse Unit 6A
glitterati Unit 6A
gorgeous Unit 6A
gown Unit 5A
gradual Unit 7B
grid Unit 3B
guild Unit 3A
harsh Unit 2A

H

heart-on-sleeve Unit 8A
heirloom Unit 5A
heyday Unit 3A
hide Unit 1B
high-end Unit 8B
highlight Unit 6B
high-profile Unit 7A
hip Unit 6B
hippie Unit 6B
horizontal Unit 1A
hover Unit 4A
hub Unit 3B
hue Unit 5A
hygroscopic Unit 2A

hygroscopicity Unit 2A
hypoallergenic Unit 2B
hypothesize Unit 1B

I

icon Unit 6B
iconic Unit 5A
identification Unit 1A
identity Unit 6A
illustrate Unit 4A
imitate Unit 8A
immense Unit 4B
imminent Unit 6B
impressive Unit 6B
inconceivable Unit 8A
incorporate Unit 2B
incredibly Unit 1A
in-depth Unit 7B
indispensable Unit 1A
individual Unit 6B
indulge Unit 4B
industrial Unit 2A
inexhaustible Unit 4B
infinite Unit 5B
influencer Unit 6B
influential Unit 6A
ingenious Unit 2B
ingrain Unit 1A
inhabitant Unit 3A
inherent Unit 2B
inherently Unit 8A
initially Unit 4A
initiate Unit 5A
initiative Unit 4A
innovation Unit 3B
innovative Unit 3A
input Unit 8A
insight Unit 7A
inspiration Unit 4A
inspire Unit 6B
insulation Unit 2A
interface Unit 8B
interior Unit 2B
interlace Unit 1A
intermarriage Unit 3B
intricacy Unit 1A
intricate Unit 5A
intrigue Unit 3A
issue Unit 6A
ivory Unit 5A

J

journalism Unit 6A
jute Unit 2A
juxtapose Unit 4A

K

khaki Unit 6B
knack Unit 1A

L

lace Unit 1B
lacquer Unit 6A
landmark Unit 7B
lappet Unit 3B
launch Unit 7B
launder Unit 5A
lavish Unit 3A

layered	Unit 5A
layette	Unit 3A
layman	Unit 8A
legacy	Unit 4B
leverage	Unit 8B
lice	Unit 1B
lignin	Unit 2A
linen	Unit 1B
lingerie	Unit 4B
lipid	Unit 2A
logistics	Unit 8B
loop	Unit 3B
lucrative	Unit 1A
luster	Unit 2A
luxurious	Unit 5A
luxury	Unit 7B

M

macramé	Unit 3A
magnificence	Unit 6A
maintain	Unit 5B
manifestation	Unit 6A
manipulate	Unit 6A
manufacture	Unit 1A
manufacturer	Unit 1B
massive	Unit 7A
matrice	Unit 2B
mature	Unit 7B
maximise	Unit 8B
medieval	Unit 3A
mesh	Unit 3A
metropolitan	Unit 4B
mighty	Unit 6B
migrate	Unit 1B
migration	Unit 1B
millennium	Unit 1A
mimic	Unit 3A
minimalism	Unit 8B
mission	Unit 7B
mite	Unit 2A
monarch	Unit 5A
monogram	Unit 3A
moribund	Unit 4B
motif	Unit 3A
motivate	Unit 6B
mourn	Unit 5A
multiple	Unit 3B
multitude	Unit 6B
mutually	Unit 8A

N

necessarily	Unit 6B
needle	Unit 3B
negate	Unit 1A
neural	Unit 8A
niche	Unit 6B
nobility	Unit 5B
nobleman	Unit 3B
non-toxicity	Unit 2A
norm	Unit 5A
nostalgia	Unit 4B
novel	Unit 8A
nurture	Unit 4B

O

obscure	Unit 3A
occur	Unit 1B

offset Unit 7A
omni-channel Unit 7B
openwork Unit 3B
optimal Unit 8B
optimism Unit 4A
optimise Unit 8B
organic Unit 2B
organza Unit 3A
original Unit 6B
originate Unit 3A
outfit Unit 1A
outlast Unit 2B
outlet Unit 3B
outperform Unit 8A
output Unit 8A
overcoat Unit 5B
overly Unit 4A
oversee Unit 4B

P

paintbrush Unit 8A
parchment Unit 3B
parliament Unit 2B
partnership Unit 8B
passion Unit 6A
patchwork Unit 4A
pattern Unit 5A
pavilion Unit 4A
pearl Unit 5A
peculiar Unit 6B
perceive Unit 7B
perception Unit 8A
perishable Unit 3A
persistently Unit 5B
personality Unit 6A
perspective Unit 6B
petticoat Unit 5A
picot Unit 3B
pictorial Unit 3B
pioneer Unit 6A
pivotal Unit 6A
plait Unit 3B
platform Unit 7B
playfulness Unit 6A
pliable Unit 3A
poignant Unit 5A
politician Unit 5B
polyester Unit 2A
polymer Unit 2A
popularity Unit 5A
porous Unit 2B
portfolio Unit 7A
portrait Unit 3B
portray Unit 6A
portrayal Unit 5A
position Unit 7B
potential Unit 2B
precious Unit 3A
predecessor Unit 5B
predict Unit 8A
preference Unit 8B
prehistory Unit 1A
premier Unit 2B
present Unit 6B
prestige Unit 7A

prestigious	Unit 4B	refine	Unit 1A
prevail	Unit 5B	reflect	Unit 6A
preview	Unit 6A	refugee	Unit 3B
previous	Unit 1B	regenerate	Unit 2A
primary	Unit 3B	rehabilitate	Unit 4B
primitive	Unit 3A	reinforcement	Unit 2A
prior	Unit 8B	religious	Unit 5B
profess	Unit 4B	remarkable	Unit 4B
professional	Unit 6B	remould	Unit 4B
proficiency	Unit 8A	render	Unit 1A
profit	Unit 7A	repeal	Unit 1A
profound	Unit 3B	repertoire	Unit 1A
progressive	Unit 4A	replacement	Unit 2A
propel	Unit 4B	represent	Unit 7B
property	Unit 2A	reputational	Unit 7A
proportion	Unit 3A	reservoir	Unit 7A
prosperous	Unit 5A	respectable	Unit 5B
prototype	Unit 2B	respectively	Unit 7A
pump	Unit 4B	restriction	Unit 4A
punk	Unit 6A	retail	Unit 7B
purchase	Unit 8B	retailer	Unit 8B
purse	Unit 8B	retain	Unit 2B
R		reveal	Unit 7A
radical	Unit 4A	revenue	Unit 4B
rapacious	Unit 7A	revitalisation	Unit 7A
rapidly	Unit 7B	revolutionary	Unit 8A
rayon	Unit 2A	ribbon	Unit 3A
reason	Unit 5A	rivalry	Unit 4B
reasonable	Unit 7B	romantic	Unit 5A
recreate	Unit 6B	rosy	Unit 7A
recyclability	Unit 2B	royal	Unit 5A
reference	Unit 3B	royalty	Unit 7A

ruff Unit 5B
runway Unit 5A

S

salient Unit 8A
sash Unit 3B
scale Unit 1B
scorpion Unit 2A
scrape Unit 1B
screenshot Unit 8B
sculpture Unit 3A
seam Unit 1B
secure Unit 4A
segment Unit 7A
seminal Unit 4A
sensibility Unit 4A
sensitive Unit 2B
sensuousness Unit 2A
sequencing Unit 1B
seriousness Unit 5B
session Unit 1B
setback Unit 7A
sewer Unit 1B
sharply Unit 7A
shelter Unit 1B
showcase Unit 4A
showy Unit 5B
signature Unit 6A
significance Unit 2B
signify Unit 5A
silhouette Unit 3A
silkworm Unit 2A
simulate Unit 8A
simultaneously Unit 7A
skepticism Unit 8A
sketch Unit 4B
slimline Unit 4B
smock Unit 3B
smuggle Unit 3B
solidify Unit 7A
sort Unit 6B
spike Unit 8B
spin Unit 1A
spinneret Unit 2A
splash Unit 6B
splendour Unit 5B
split Unit 1B
spokesperson Unit 6B
sponsorship Unit 7A
sportswear Unit 7A
spot Unit 8B
stalk Unit 4A
staple Unit 1A
statue Unit 1B
steadfast Unit 4A
steadily Unit 5B
stewardship Unit 4B
stock Unit 7A
strategic Unit 7B
strategy Unit 7A
streamline Unit 1A
strengthen Unit 2B
stubborn Unit 4B
sturdy Unit 3A
stylish Unit 5B

substantial	Unit 3A	transition	Unit 6B
substitute	Unit 2A	trendy	Unit 6B
sumptuous	Unit 5B	trim	Unit 5A
superiority	Unit 2B	trimming	Unit 3B
supersede	Unit 2B	tulle	Unit 5A
supplement	Unit 4A	turnaround	Unit 5B
surge	Unit 6B	tuxedo	Unit 5B
surpass	Unit 2A	tweak	Unit 4B
suspicion	Unit 8A	tweed	Unit 4B
sustainability	Unit 2A	twig	Unit 1A
symbolic	Unit 6A	**U**	
symbolically	Unit 8A	ultimately	Unit 6A
symbolism	Unit 5A	ultraviolet	Unit 2B
synthesis	Unit 2A	unconsciously	Unit 6A
synthetic	Unit 2A	unconventional	Unit 8A
T		undeniably	Unit 6A
tactic	Unit 7A	undergarment	Unit 2A
target	Unit 7B	underlie	Unit 1A
technique	Unit 3A	underscore	Unit 1A
technological	Unit 1B	undisputed	Unit 5B
temporary	Unit 3B	unique	Unit 4A
testimony	Unit 4B	unravel	Unit 3B
textile	Unit 1B	unsettling	Unit 8A
texture	Unit 3A	upgrade	Unit 7B
thereby	Unit 5B	upheaval	Unit 1B
thermal	Unit 2A	utility	Unit 6A
thermoplastic	Unit 2A	**V**	
thread	Unit 3B	valid	Unit 5B
thrive	Unit 1A	validity	Unit 8B
transform	Unit 4A	valuable	Unit 2A
transformation	Unit 7B	variety	Unit 1B
transit	Unit 8B	various	Unit 6B

vellum Unit 3A
verbally Unit 6A
versatile Unit 2B
vertical Unit 1A
vest Unit 5B
via Unit 6A
viable Unit 8A
violet Unit 5A
virtually Unit 3B
virtuosity Unit 4B
vital Unit 6A
vivid Unit 7B
vogue Unit 5B
voracious Unit 4B

W

wane Unit 3A
wardrobe Unit 1B
wastage Unit 8B
weaken Unit 2B
weave Unit 1B
weft Unit 1A
well-received Unit 8A
well-rounded Unit 7B
wig Unit 5B
withstand Unit 2B
wool Unit 3A
wrap Unit 1A

Y

yarn Unit 1A
youthful Unit 7B

Z

Zeitgeist Unit 3A